Collins

AF778451

White Rose Maths

White Rose Maths
Key Stage 3
Behind the Questions
Teacher Guide 3

Ian Davies, Caroline Hamilton and Sahar Shillabeer

William Collins' dream of knowledge for all began with the publication of his first book in 1819. A self-educated mill worker, he not only enriched millions of lives, but also founded a flourishing publishing house. Today, staying true to this spirit, Collins books are packed with inspiration, innovation and practical expertise.
They place you at the centre of a world of possibility and give you exactly what you need to explore it.

Collins. Freedom to teach.

Published by Collins
An imprint of HarperCollins*Publishers*
The News Building
1 London Bridge Street
London
SE1 9GF

1st Floor, Watermarque Building
Ringsend Road
Dublin 4
Ireland

Browse the complete Collins catalogue at www.collins.co.uk

10 9 8 7 6 5 4 3 2 1

ISBN: 978-0-00-840093-4

British Library Cataloguing-in-Publication Data
A catalogue record for this publication is available from the British Library.

Authors: Ian Davies, Caroline Hamilton and Sahar Shillabeer
Series editors: Ian Davies and Caroline Hamilton
Publisher: Katie Sergeant
Product Manager: Jennifer Hall
Product developer: Natasha Paul
Content editor: Tina Pietron
Editors: Karl Warsi, Julie Bond, Phil Gallagher, Tim Jackson
Proofreader: Laurice Suess
Project Manager: Karen Williams
Cover designer: Kneath Associates Ltd
Internal designer and illustrator: Ken Vail Graphic Design, Ltd
Typesetter: Ken Vail Graphic Design, Ltd
Production controller: Katharine Willard
Printed and bound in the UK using
100% Renewable Electricity at CPI Group (UK) Ltd

Photo acknowledgements
The publishers wish to thank the following for permission to reproduce photographs. Every effort has been made to trace copyright holders and to obtain their permission for the use of copyright materials. The publishers will gladly receive any information enabling them to rectify any error or omission at the first opportunity.
p104 l topseller/Shutterstock, p104 r Catharina van Delden/Shutterstock.

FSC www.fsc.org
MIX
Paper from responsible sources
FSC™ C007454

This book is produced from independently certified FSC™ paper to ensure responsible forest management.

For more information visit: **www.harpercollins.co.uk/green**

Acknowledgements
The publishers gratefully acknowledge the permission granted to reproduce the copyright material in this book. Every effort has been made to trace copyright holders and to obtain their permission for the use of copyright material. The publishers will gladly receive any information enabling them to rectify any error or omission at the first opportunity.

Contents

Introduction: Behind the questions

The White Rose Maths approach

The White Rose Maths textbook series guides students and teachers through a comprehensive learning journey to help build a culture of deep understanding and competence in mathematics. It fits perfectly alongside the White Rose Maths Secondary schemes of learning but can also be used alongside other existing materials or curricula. The books support a mastery approach to teaching and learning and have been designed to support the aims and objectives of the Key Stage 3 National Curriculum. This states that all students should:

- become **fluent** in the fundamentals of mathematics, including through varied and frequent practice with increasingly complex problems over time, so that pupils develop conceptual understanding and the ability to recall and apply knowledge rapidly and accurately
- **reason** mathematically by following a line of enquiry, conjecturing relationships and generalisations, and developing an argument, justification or proof using mathematical language
- **solve problems** by applying their mathematics to a variety of routine and non-routine problems with increasing sophistication, including breaking down problems into a series of simpler steps and persevering in seeking solutions.

We recognise that time is an issue and it's not easy to cover the full content of the curriculum in the depth you would wish to. These books are designed to give time to think about a topic, to develop understanding and also to realise that mastery will not have been fully achieved by the end of a chapter, whatever its length. Our small step approach has been devised to ensure that students will return to a topic time and time again, both within the further study of that topic and also in other areas of mathematics; this iterative process of revisiting and interleaving will enable students to deepen their understanding.

A mastery approach

Helping students on their journey towards mastery is exactly what White Rose Maths exists to do. It isn't always an easy journey, and it's important to realise that there's more than one way to get there. We think mastery of anything – playing an instrument, speaking a new language, mathematics – takes a very long time. That's why we talk about the "journey to mastery", rather than "having mastered". For example, children start learning to add in early years and keep developing their skills over many years – from single-digit numbers, to multi-digit numbers, then decimals, then fractions, then negatives, addition in different units, and so on. It would take several years to fully master addition, arguably one of the most basic concepts in mathematics, so what we do is break the journey down into small steps, spending time carefully considering each step. Through intelligent practice and building up experience of different contexts, gradually we move towards mastery where students are fluent in the unfamiliar and can apply their skills in any new situation.

A concrete-pictorial-abstract approach is an important aspect of the mastery approach and is based on evidence and research. Some of this research comes from Jerome Bruner. In his research on the cognitive development of children (1966), he proposed three modes of representation:

- **Enactive representation (concrete)** – students should have the opportunity to use concrete objects and manipulatives to help them understand what they are doing.
- **Iconic representation (pictorial)** – alongside manipulatives, students can use pictorial representations. These representations can then be used to aid reasoning and problem solving.
- **Symbolic representation (abstract)** – both concrete and pictorial representations should support students' understanding of abstract methods.

The White Rose Maths textbook series is unique in that it has concrete and pictorial representations at its heart. Powerful tools such as bar models and double-sided counters will provide many lightbulb moments for your students.

Sequencing of content

Book 3 follows the same rationale as Books 1 and 2. Students have built a good understanding of algebraic notation and its applications in the first two books and Book 3 deepens this by looking at relationships more formally, developing students' understanding of proof. Proportional reasoning is again emphasised and returned to several times throughout the course of the book with students increasing their knowledge of different kinds of relationships, including inverse proportion.

Much of the other content in this book also builds on students' learning in Books 1 and 2. It is not assumed that students will have remembered everything from their earlier studies and when necessary prior learning is reviewed before moving on to the newer content. Opportunities to revisit learning in other contexts are also included throughout the book. For example, forming and solving equations and inequalities is included in many blocks.

The books are structured to support students in performing simpler tasks, so they can then move on to more complex tasks. For example, when exploring rates of change and their units in Block 14 we begin with the everyday familiar notion of speed but moving on to the more abstract notion of density and then generalising to look at any type of rate.

The small steps progression from the White Rose Maths scheme of learning form the sequencing of concepts within each block. Take a look at this example from the Testing Conjectures block.

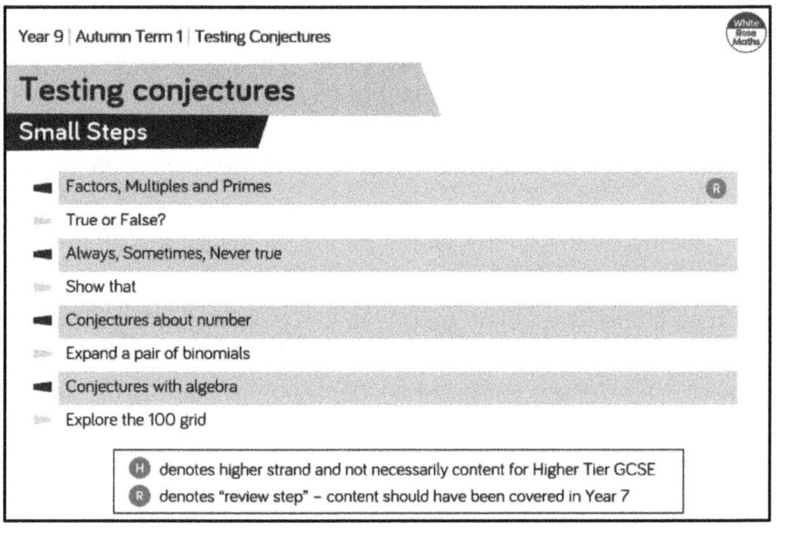

As you can see, nothing is left to chance. The learning is broken down carefully to ensure that students have the appropriate building blocks to access all of the content. We begin by revising factors, multiples and primes as these key concepts will provide a rich basis for exploring and testing conjectures. Students firstly check whether given statements are true or false, learning about examples and counterexamples. This is developed slightly to then look at statements that may be always, sometimes or never true, appreciating that some examples may support and others may not support a conjecture. Gradually we build up to more formal algebraic proof, supporting this with the same concrete and pictorial representations students have used earlier to develop their understanding of algebraic notation.

The expectation is that most students will study Book 3 in Year 9, the final year of Key Stage 3. The White Rose Maths schemes of learning allow time for revision, consolidation and extension before moving on to Key Stage 4 in Year 10. To support this, Block 17 of Book 3 includes three suggested areas of study

- Handling Data Review. There is no new Handling Data in Book 3, so this block could be used for revision or to learn about finding the mean from frequency tables if these were not covered as Higher steps in Book 2
- Sequences Review. Likewise this provides an opportunity to revise all types of sequence and how to the rule for the *n*th term of a linear sequence which was a Higher step in Book 2
- Moving on with trigonometry. This develops the learning in Block 12 of Book 3 to explore general angles. Trigonometry is studied in depth at the start of the White Rose Maths schemes of learning for Key Stage 4

As well as to revise, all these chapters include a series of challenges to extend students' learning.

White Rose Maths Key Stage 3 textbook series components

Student book

There is one student book per year for KS3: Book 1, Book 2 and Book 3, corresponding to Years 7, 8 and 9 in the White Rose Maths schemes of learning. Each book caters for all attainment levels, and would work well for either setted or mixed-attainment groups. Each block of learning consists of:

- **A block opener** – This gives a visual of the topics that will be studied in the block and the key representations that will be used.
- **Between two and eight chapters** – The number of chapters per block varies depending on the amount of content covered. Each chapter comprises:
  - A list of the White Rose Maths small steps that will be covered
  - The key words used in the chapter
  - An "Are you ready?" section – this is a short set of questions to check that students possess the prior knowledge needed to access the new learning.
  - Visuals and accompanying explanation of the models and images used in the chapter, such as bar models or place value charts.
  - Examples and practice – these are repeated in up to three sections: Practice A, Practice B and Practice C. The number of practice sections per chapter varies depending on how much content is being covered. Each practice section involves fluency, reasoning and problem-solving questions to ensure all students cover the main aims of the National Curriculum. Practice is often scaffolded with diagrams to help students to "see the maths". You will notice symbols are used throughout to prompt students to use concrete manipulatives, draw a diagram or discuss their thinking with others (these icons are fully explained in the Introduction at the start of the student book). These are all important aspects of a mastery approach to teaching and learning in mathematics.
  - 'What do you think?' sections – located at the end of each practice section, these questions are often more open-ended, or encourage deeper reasoning and problem-solving. We encourage all students to attempt these questions with support from whole-class or paired discussion. Students should expect to have to think hard or think differently before answering.
  - A "Consolidation" and a "Stretch" section – towards the end of the chapter your class is likely to diverge. They will either complete the Consolidation section or the Stretch section depending on how they found the practice questions. Consolidation gives students a chance to attempt some more of the fluency-style questions. These are really useful if they have needed help throughout the main practice, as they could now attempt similar questions independently. Consolidation could also be used for revision, coming back to the topic at a later date, perhaps after assessment has suggested a gap in knowledge. Students who are confident with the practice questions get the chance to challenge themselves in the Stretch section. These questions often bring together numerous mathematical ideas and include rich problem-solving tasks.
  - A time to reflect – finally, students are asked to reflect on their learning, which may include putting key ideas into their own words, or justifying why something is or isn't true.
- **A "Check my understanding" quiz** – Each block of learning ends with a short quiz covering content from all the chapters. Prior to attempting the quiz, students can read the statements about what they should have become fluent in and how they have used their reasoning and problem-solving skills. The questions in the quiz will indicate to both students and teachers the extent of retention of the key aspects of the block, and whether students need further revision or consolidation.

You will notice that we use an "H" symbol at times throughout the student books. This denotes content to stretch and challenge students. It does not necessarily mean that it is Higher content at GCSE; for example, in Book 2, constructing bisectors is Higher content but appears on the Foundation tier of GCSE. Use these chapters with as many students as possible; however, do not worry if you miss them out. The Higher content at

KS3 is covered again in later years, when it might be more appropriate for some of your students. The "R" symbol in Books 2 and 3 denotes review content. This material has been covered previously, and this is a chance to revise and extend students' learning in these areas.

Behind the questions

A teacher guide is provided for each student book. The aim is to convey the rationale behind each chapter and give teachers an insight into the authors' thinking when formulating questions or sets of questions. You will find that conceptual and procedural variation is used throughout the books. The teacher guide helps you see when and why this is used.

Each student book chapter has a corresponding "Behind the questions" section. This includes:

- The small steps covered from the White Rose Maths scheme of learning.
- The rationale for the learning taking place – why is it important, why is it taught in this place and what prior learning does it build on?
- Things to look out for – this section includes the common misconceptions or sticking points for students when learning this content.
- Interleaving opportunities – interleaving is an important aspect of successful learning and retention in mathematics. It is important that students see mathematics as an interconnected web of ideas and concepts, not stand-alone topics that have no relationship with each other. Interleaving not only helps students build connections but also gives the opportunity to revisit content and further consolidate it.
- Models and representations – a reminder of some of the key models and representations from the chapter and a rationale for why they are used.
- Commentary for the questions used in each section of the chapter – here, the authors choose some key questions and explain their thinking behind them. Often numbers will have been chosen very carefully to draw out a specific point. Reading this will help you, as the teacher, to facilitate rich discussion and use careful questioning to draw out big ideas.

You can download editable versions of these teacher materials from collins.co.uk/WhiteRoseMaths/downloads

Additional materials and training available

White Rose Maths produce a great deal of extra material, often free, that would work well alongside this textbook series. They have a free, low-stakes assessment for each block of learning available plus a comprehensive set of more formal end-of-term assessments. These can all be accessed from their website www.whiterosemaths.com.

In addition, if you would like some training on some of the approaches used in the books, this can be booked either online or face-to-face. Top recommendations would be the courses entitled *Concrete-Pictorial-Abstract Methods*, *An Introduction to Bar Modelling*, *Mathematical Talk and Questioning*, *Reasoning and Problem Solving*, and *Thinking Through Variation*.

A note from the series editors

We are extremely excited by the release of this textbook series. Our passion is to help as many children and young people as we can to enjoy and understand mathematics. We hope that you can see this passion come through in our carefully thought-through questions, our use of models and images to aid understanding, and our interesting and rich reasoning and problem-solving tasks.

To us this is more than a textbook, it is a book full of exciting mathematical connections ready to inspire and enthuse teachers, students and parents alike, and we are incredibly proud to have joined the Collins team in the journey of its creation. We hope you enjoy using it as much as we enjoyed writing it.

Ian Davies (Head of Curriculum, White Rose Maths) **Caroline Hamilton** (Head of White Rose Maths)

Overview of course

Here is a handy overview of the blocks covered in each of the books in the White Rose Maths Key Stage 3 scheme, showing how you could fit them into your yearly programme for mathematics teaching.

Term	Book 1	Book 2	Book 3
Autumn	1. Sequences	1. Ratio and scale	1. Straight line graphs
	2. Algebraic notation	2. Multiplicative change	2. Forming and solving equations
	3. Equality and equivalence	3. Multiplying and dividing fractions	3. Testing conjectures
	4. Place value and ordering	4. Working in the Cartesian plane	4. Three-dimensional shapes
	5. Fractions, decimals and percentages	5. Representing data	5. Constructions and congruency
		6. Tables and probability	
Spring	6. Addition and subtraction	7. Brackets, equations and inequalities	6. Numbers
	7. Multiplication and division	8. Sequences	7. Percentages
	8. Fractions and percentages of amounts	9. Indices	8. Maths and money
	9. Directed number	10. Fractions and percentages	9. Deduction
	10. Fractional thinking	11. Standard index form	10. Rotations and translations
		12. Number sense	11. Pythagoras' theorem
Summer	11. Construction and measuring	13. Angles in parallel lines and polygons	12. Enlargement and similarity
	12. Geometric reasoning	14. Area of trapezia and circles	13. Solving ratio and proportion problems
	13. Developing number sense	15. Line symmetry and reflection	14. Rates
	14. Sets and probability	16. The data handling cycle	15. Probability
	15. Primes and proof	17. Measures of location	16. Algebraic representation
			17. Preparation for KS4

1.1 Plotting and reading graphs

Small steps

- Recognise lines parallel to the axes, and also the lines $y = x$ and $y = -x$ (R)
- Use tables of values (R)

Rationale

In this chapter, students recap drawing straight-line graphs. They will have covered this topic previously in Year 8 and will build on it later in the block. It is essential that students can draw and recognise lines of the form $y = a$, $x = b$, $y = x$ and $y = -x$, and they should also be able to use a table of values to plot other lines that are given in the form $y = mx + c$. Correct mathematical language when reading the equation of a line can assist understanding. For example, reading $x = 4$ as "x is equal to 4" helps students to understand why, at every point on that line, the x-value is equal to 4.

Things to look out for

- Some students find it difficult to draw and scale the coordinate axes correctly. Consider providing printed axes for these students to avoid that barrier.
- Students may mix up the x- and y-values in a pair of coordinates, which can result in incorrect lines being drawn and incorrect points being generated. Questions such as 1 and 2 in the "Are you ready?" exercise address this problem.
- Students may confuse the axes and hence plot points incorrectly. Encourage them to always label the axes; it may be helpful to remind them that "x is a<u>cross</u>" and "y goes to the sk<u>y</u>".

Interleaving opportunities

- Students will consolidate working in all four quadrants and plotting coordinates.
- Substitution is an essential prerequisite for this chapter as students need to be able to work out coordinates of points on a line.
- Sequences are revisited in the "Are you ready?" exercise, and this can be linked to the sequence formed by the y-values on any given line.
- Students will work with algebraic notation and need to be able to recognise different forms of the same expression, e.g. $\frac{2}{3}x$ and $\frac{2x}{3}$
- Students will be required to use directed numbers in their calculations, which provides a valuable revision opportunity.

Models and representations

- Tables of values can be used to work out the coordinates of any point on a given line. In this table of values, the y-value is 3 times the x-value, so this is the table of values for $y = 3x$.

x	0	1	2	3
y	0	3	6	9

- A function machine can be used alongside a table of values to help students to calculate missing values. This function machine can be used to calculate pairs of values that lie on the line $y = 2x + 5$.

Input x → $\times 2$ → $+ 5$ → y Output

Behind the questions

Are you ready?

These four questions cover the essential prerequisite knowledge for this chapter. Questions 1 and 2 ensure that students are able to confidently differentiate between the x- and y-values in any given pair of coordinates. In question 3, students are reminded of the structure of an algebraic expression and what it represents. Questions 3 and 4 are linked: first students are asked to explain the meaning of an expression and then they go on to evaluate it for given values. This will be built on later in the chapter when substituting into equations to generate pairs of coordinates.

Practice 1.1A

1 and **2** These two questions are designed to ensure that students can recognise and are able to explain in their own words that lines of the form $y = a$ are horizontal lines that go through point a on the y-axis, and lines of the form $x = b$ are vertical lines that go through point b on the x-axis. Students could be encouraged to read off coordinates as a method of checking their answers. You could check students' understanding by asking them to give the coordinates of points that they can't see.

3 Students need to recognise that the line in part f is the y-axis; you could also ask them to give the equation of the x-axis.

4 Students are asked to find the equation of a line rather than plot it. They should know from earlier questions that a line of the form $x = b$ is a vertical line through b on the x-axis, so they should be able to transfer this knowledge and conclude that a vertical line through 18 on the x-axis is the line $x = 18$. Writing down coordinates of points that lie on a line can help to identify this.

5 Here students are reminded of the lines $y = x$ and $y = -x$. Using correct mathematical language such as reading $y = x$ as "y is equal to x" can help to students understand why, at any point on the line, the y-value is equal to the x-value. Students could be stretched to consider why these lines are not horizontal/vertical.

What do you think? 1.1A

Questions 1 and 2 ensure that students can identify the key features of the straight-line graphs they've covered so far when these are described in different ways. Question 3 makes students aware that a scale can affect the appearance of a straight line so that they don't think that $y = x$ is always at a 45° angle to the axes.

Practice 1.1B

1 and **2** These two questions are linked as they are both based on the same four one-step equations. Question 1 links the equation of a straight line to its corresponding function machine, whilst question 2 asks students to explain this link in relation to pairs of coordinates and then draw the line.

3 and **4** Again, these questions are linked to help build understanding, but this time the focus is on two-step equations. Notice the similarities between the numbers and operations used in each equation. Students should be able to use their knowledge of the order of operations to match each equation to a function machine.

5 This builds on question 4. This time the scaffolding of function machines is not given. However, students could be encouraged to draw them for support, if needed. Question 6 asks students to think about the similarities and differences.

7 to **9** are designed to make students become aware of how a negative coefficient of x affects the points plotted and the appearance of the line. They should notice from the table of values that as x increases, y decreases, and realise that this gives a line that slopes downwards to the right.

What do you think? 1.1B

These three questions build on the ideas covered in Practice 1.1B and get students to think more deeply. Question 1 asks students to consider how they can instantly recognise a line with a negative gradient; this will be built upon later in the block. Questions 2 and 3 look at sequences within y-values, but highlight that this is only an effective method when the x-values are consecutive.

Consolidate – do you need more?

This revisits the key points covered in this chapter and ensures that students can plot straight-line graphs.

Stretch – can you deepen your learning?

1 and **2** introduce the concept of the y-intercept, although this vocabulary is not explicitly used. Students should make connections between the equation of a line and its key features.

4 introduces the idea of a point of intersection and then question 5 builds on this to include a third line. Students will need to use substitution to check whether all three lines intersect at one point. Although some students may draw the lines to test this, they should then be encouraged to consider how they could do this without drawing.

Reflect

Students are encouraged to reflect on their learning in their own words and describe the relationship between the equation and the appearance of a straight-line graph.

1.2 Gradients and intercepts

Small steps

- Compare gradients
- Compare intercepts
- Understand and use $y = mx + c$

Rationale

Although students may be familiar with the gradient and intercept of a straight-line graph and may have started to spot patterns or make predictions, here they explore these concepts in more detail. It is essential that they can articulate that the gradient of a straight line is given by the coefficient of x and that the y-intercept is given by the constant in the equation. Students should recognise that lines with the same gradient are parallel and that lines with the same y-intercept will cross the y-axis at the same point. They should be reminded to pay close attention to whether they are asked for the value of the y-intercept or the coordinates of it. They should recognise that the coordinates of the y-intercept will always be of the form $(0, a)$ because it is a point on the y-axis, which is also the line $x = 0$.

Things to look out for

- Some students find it difficult to draw and scale the coordinate axes accurately. Consider providing printed axes for these students.
- Students might generalise that the gradient is always the first number after the "equal to" symbol, which can cause errors if an equation is not given in the form $y = mx + c$. Instead, they should be encouraged to describe it as the coefficient of x. Similar issues can arise with the y-intercept.
- Overlooking whether the gradient/intercept is positive or negative can cause students to think that $y = 2x$ and $y = -2x$, for example, are parallel. They need to fully understand the difference between lines with positive and negative gradients.
- Once students have made a connection between the gradient and the difference between successive y-values in a table of values, they can rely on this too strongly and will make mistakes when the x-values are not consecutive.

Interleaving opportunities

- Students will consolidate working in all four quadrants and plotting coordinates.
- Substitution is an essential prerequisite for this chapter as students need to be able to work out coordinates of points on a line.
- Sequences are revisited in the "Are you ready?" exercise, and this can be linked to the sequence formed by the y-values on any given line.
- Students will work with algebraic notation and need to be able to recognise different forms of the same expression, e.g. $\frac{2}{3}x$ and $\frac{2x}{3}$
- Students will be required to use directed numbers in their calculations, which provides a valuable revision opportunity.

Models and representations

- Tables of values can be used to work out the coordinates of any point on a given line. In this table of values, the y-value is 3 times the x-value, so this is the table of values for $y = 3x$.

x	0	1	2	3
y	0	3	6	9

- A function machine can be used alongside a table of values to help students to calculate missing values. This function machine can be used to calculate pairs of values that lie on the line $y = 2x + 5$.

Input: x → $\times 2$ → $+ 5$ → y :Output

Behind the questions

Are you ready?

These three questions cover the essential prerequisite knowledge for this chapter. Questions 1 and 2 revisit learning from the previous chapter and ensure that students can generate points given an equation and use these to draw a line. Question 3 revisits earlier learning on sequences as this can be linked to why, for example, the y-values on the line $y = 3x + 1$ increase by 3 as x increases by 1.

Practice 1.2A

1 and **2** Here students focus solely on the gradient of a line and how this affects the appearance of the line. They should be able to articulate that as the gradient increases, the line becomes steeper. In question 2, this could be described as being "steeper downhill". Students should recognise that when the gradient is positive the line slopes upwards, and when it is negative it slopes downwards.

3 ensures that students understand the difference between a positive gradient and a negative gradient. This will be invaluable in later learning when sense-checking any drawings or equations of lines.

4 ensures that students can identify parallel lines both from a graph and from the equations of the lines. They should recognise that lines with the same gradient are parallel.

5 to **8** These four questions take a similar approach to that of questions 1 to 4 but this time the focus is on the y-intercept rather than the gradient. Notice how the lines sketched in question 7 are identical to those in question 3, but here students are focusing on a different feature of the line. Students could be stretched to suggest possible equations of each line.

What do you think? 1.2A

These four questions require students to think more deeply about gradients and intercepts. Question 1 addresses the common misconception that the gradient is the first number after the "equal to" symbol. In question 3, students should recognise that there are several possible answers for part b and be able to explain why. They could also be asked to consider the limitations on the range of possible answers.

Practice 1.2B

1 to **3** Here students consolidate interpreting the gradient and y-intercept of a line given its equation. In question 1, all equations are given in the form $y = mx + c$. In question 2, their attention is drawn to the sign attached to each number and what this means. In question 3, the equations are not given in the form $y = mx + c$, so students need to pay close attention to what each part of the equation represents.

4 focuses on the idea of lines being parallel or having the same y-intercept. It also provides the opportunity for students to consolidate fraction and decimal equivalence.

5 Students should recognise that there are an infinite number of possibilities for each part, and should be able to identify the similarities and differences between all these possibilities.

8 to **10** Here students should recognise that there is only one possible solution for each part and should be able to articulate why this is the case.

What do you think? 1.2B

These four questions require students to think more deeply about gradients and intercepts. Questions 1 to 3 explore common misconceptions, and students should be encouraged to explain the mistakes that have been made as well as give the correct answers.

Consolidate – do you need more?

The key points covered in this chapter are revisited to ensure that students can plot straight-line graphs.

Stretch – can you deepen your learning?

1 Students should recognise that these are the same line and so they are not parallel.

2 Students may not realise that they are able to find the coordinates of point R, and so may need guidance.

4 Students should be encouraged to first form an equation to find the given values.

6 formally explores the idea of there being an infinite number of possibilities for equations of parallel lines.

Reflect

Students reflect on their learning and explain how to identify the gradient and y-intercept of a line given its equation.

1.3 Equations of lines

Small steps

- Write an equation in the form $y = mx + c$ (H)
- Find the equation of a line from a graph
- Interpret gradient and intercepts of real-life graphs

Rationale

In this chapter, students further explore equations of straight lines. In the first part of the chapter, they look at lines whose equations are not given in the form $y = mx + c$. Note that, at this point, students have not yet learnt how to rearrange equations. For this reason, the questions and examples given are all based around lines of the form $y = mx + c$, $y - c = mx$ or $y - mx = c$. The focus is not on teaching students how to rearrange equations but, instead, it is on recognising that the equation of the line is not given in its usual form and so some form of adjustment is needed. The rearrangement required is basic and involves only one step.

In the second part, students build on their learning from the previous chapter and identify the equation of a straight line from its graph. Ask questions such as "Is the gradient/y-intercept positive or negative? How do you know?" to enable students to sense-check their answers. These concepts are then built on by looking at real-life graphs that require students to interpret both the gradient and the vertical axis intercept in context.

Things to look out for

- Not recognising when the equation of a line isn't given in the form $y = mx + c$. Students could be encouraged to substitute values into their equations to help to recognise any mistakes.
- Where the equation of a line is in the form $ay = amx + ac$, some students might divide only one side of the equation by a. Remind them of the meaning of equality to overcome this.
- Students may incorrectly calculate the gradient of a line by not acknowledging which way it is sloping. Asking them beforehand to identify whether the gradient is positive or negative can help to overcome this.
- When a graph relates to a context, students may disregard this and just give numerical answers. They should be prompted to look carefully at the information being represented and to interpret their answers in this context.

Interleaving opportunities

- Students will need to use their knowledge of number sense and related facts. This will enable them to rewrite equations in the form $y = mx + c$ without any formal knowledge of rearranging equations.
- Students will revisit their learning from the previous chapters in the block, in particular the content from Chapter 1.2 relating to the sign of the gradient/y-intercept of a straight line.
- Students calculated the gradient of a line when learning about ratio and scale in Book 2. Some students might need to consolidate their understanding of this before moving on to finding the equation of a line.
- Substitution should be used as a method of checking answers throughout.

Models and representations

Coordinate grids are used throughout. Remind students to pay close attention to the scale on each axis rather than assuming that the numbers will go up in 1s.

Behind the questions

Are you ready?

These four questions revisit the essential prerequisite knowledge for this chapter. Question 1 looks at related number facts which will be built upon when writing an equation in the form $y = mx + c$. Questions 2 and 3 revisit learning from previous chapters and question 4 revisits proportional reasoning from Book 2, which links to real-life graphs later in the chapter.

Practice 1.3A

1 and **2** are based on the same equations. In question 1 the focus is simply on writing each equation in the form $y = mx + c$ whilst, in question 2, students go on to interpret these equations.

3 prompts students to recognise that if the constant is on the opposite side of the "equal to" symbol to x, then they can't simply read this off as the y-intercept.

4 and **5** are similar to questions 1 and 2 but follow on from question 3. Here, students are required to perform a one-step rearrangement so that the equations of the lines are in the correct form. In question 5, they also need to interpret these equations.

6 and **7** So far, students have been focusing on one skill at a time, but here they are required to decide what steps need to be taken to write the equation in the form $y = mx + c$ and then interpret the equations.

What do you think? 1.3A

These three questions require students to think more deeply about writing equations in the form $y = mx + c$. Questions 1 and 2 draw attention to the fact that the equation of a line can be represented in several different ways. In question 2, students could be stretched to consider another set of cards that belongs in each of the given sets.

Practice 1.3B

1 Here the focus is solely on students reading off the y-intercept of each line and then writing its equation. Students could be prompted to recognise that, for example, in part a, the line will be of the form $y = x -$ ___ before reading off the value and explaining why this is the case.

2 revisits learning from Book 2 on finding the gradient of a line. Students should know that the gradient is how many units are moved up/down for each unit moved across. The units here are squares.

3 Notice that the lines have the same gradients as those in question 2 but, in this question, they are drawn on coordinate grids so that students are able to find their equations.

4 and **5** provide further practice at finding the equation of a straight line from its graph.

6 to **8** look at straight-line graphs in real-life contexts. Students should be able to interpret the gradient and y-intercept in the context of the question.

What do you think? 1.3B

These questions require students to think more deeply about finding the equation of a line from a graph. Questions 1 and 2 address common misconceptions whilst question 3 interleaves learning from the previous chapter.

Consolidate – do you need more?

These five questions revisit the key points covered in this chapter and ensure that students can write the equation of a straight line in the form $y = mx + c$ and also find the equation of a straight line from its graph.

Stretch – can you deepen your learning?

2 looks in more detail at lines drawn on axes with different scales. Students should recognise that the same line has been drawn in all four parts and consider how the different scales affect the equation.

3 Here students are required to add scales to the axes before they can work out the equation of the line.

4 introduces the idea of finding the equation of a line that passes through two points. Students are encouraged to draw a diagram for support.

Reflect

Students are encouraged to identify what they think are the most common misconceptions with this topic. By exploring them and addressing them, students are less likely to make these mistakes themselves.

1.4 Further graphs (H)

Small steps

- Explore perpendicular lines (H)
- Model real-life graphs involving inverse proportion (H)

Rationale

This chapter covers some Higher content of the block on straight-line graphs. In the first part of the chapter, students explore perpendicular lines. They should be told that when two lines are perpendicular, their gradients have a product of −1 and then use this, along with their knowledge of gradients from the previous chapters, to identify perpendicular lines. Students should already be familiar with the concept of a reciprocal from their work on dividing fractions in Book 2 but might need reminding of this, which is why it is included in the "Are you ready?" section.

In the second part of the chapter, students are introduced to the idea of inverse proportion. All their work on multiplicative reasoning so far has looked at quantities that increase at the same rate; here they are exposed to examples where this isn't the case. You could use real-life ideas to support their understanding. For example, in a non-mathematical context, students should appreciate that if more cleaners are working, it will take less time to do the cleaning. They can then be introduced to the maths.

Things to look out for

- Just finding the reciprocal rather than the negative reciprocal when working with perpendicular lines. Students could be encouraged to sketch each line to avoid making mistakes.
- Some students might treat inverse proportion in the same way as direct proportion because they're more confident when working with this type of relationship. Students should be encouraged to sense-check their answers and consider whether or not they are realistic in the context of the question.

Interleaving opportunities

- Students will make use of their prior learning on multiplying and dividing fractions when working with reciprocals in this chapter.
- Students are use their learning from the earlier chapters in this block when interpreting gradients and intercepts from graphs.
- This is a good opportunity to revisit linear and non-linear relationships; students should appreciate that an inversely proportional relationship is non-linear.

Models and representations

- Tables of values are used to promote understanding of inversely proportional relationships. Where *x* and *y* are inversely proportional, the product of *x* and *y* is equal for every pair of values.

x	1	2	5	10	100
y	100	50	20	10	1

Behind the questions

Are you ready?

These three questions cover the essential prerequisite knowledge for this chapter. Question 1 looks at area, which will be revisited when working with inverse proportion in Block 13. The length and width of a rectangle are inversely proportional when the area of the rectangle is constant. Questions 2 and 3 remind students of the idea of a reciprocal; they should know that the reciprocal of a number is its multiplicative inverse.

Practice 1.4A

1 Here students consolidate their learning from the previous chapter on finding the gradient of a line from its graph. They should be encouraged to spot links and make connections so as to recognise that when two lines are perpendicular, the gradient of one line is the negative reciprocal of the gradient of the other.

2 Students complete each calculation by filling in the negative multiplicative inverse of each number.

3 Here students are asked to calculate the negative reciprocal of each given number.

4 Students combine the skills used in questions 1 to 3 by writing the equation of a line perpendicular to each of the given lines. Notice that each perpendicular line goes through the origin so students only need to focus on the gradients and not the y-intercepts.

5 Here students need to rewrite each equation in the form $y = mx + c$ before showing that they are perpendicular.

6 Unlike question 4, in question 6 there are an infinite number of possible solutions, whilst in questions 7 and 8 there is exactly one solution for each question; students should be able to articulate why that is the case.

What do you think? 1.4A

These three questions require students to think more deeply about perpendicular lines, with a focus on fractional and decimal equivalents. Students should be encouraged to write any decimals as fractions before calculating the negative reciprocal.

Practice 1.4B

1 to **4** Here the contexts are chosen so that students should be able to link them to real-life situations and sense-check their answers. This should help to prevent them incorrectly using a multiplicative relationship. In question 1, all the minutes are integer values. In question 2, students need to work with decimal values and, in question 3, the decimal values become a little harder to spot. By this point in the exercise, students should recognise that the product of the pair of values in every column is the same.

4 Here students should be encouraged to first draw the table of values to support their thinking.

What do you think? 1.4B

These three questions require students to think more deeply about inverse proportion. For question 1, they should use common sense to consider each real-life context. For example, if there are no bakers, the cakes will never get iced. If students have recognised that the product of the values in each column of the table of values is constant, then they should also recognise that none of the values can be 0 or they couldn't get the desired product. Question 2 highlights that inverse proportion is not linear whilst question 3 draws students' attention to a relationship that is neither directly nor inversely proportional: the number of ovens doesn't affect the cooking time.

Consolidate – do you need more?

These four questions revisit the key points covered in this chapter and ensure that students can work with perpendicular lines and basic inverse proportion. Ensure that students appreciate any links drawn between gradient and negative reciprocals.

Stretch – can you deepen your learning?

1 to **3** require students to use their knowledge of algebraic notation in the context of perpendicular lines and inverse proportion. Students could be stretched to consider the limitations on the variables in each question.

4 looks at the idea of finding the equation of a perpendicular line that passes through a given point. Students should first find the gradient of the perpendicular line before considering the relationship between the x- and y-values at the given point and using this to identify the y-intercept.

5 Students might need prompting to recognise that the two lines have the same x-intercept and that they can work out the coordinates of this point.

Reflect

Students reflect on their learning and explain in their own words how to tell whether two lines are perpendicular from their equations and from their graphs. They should also be able to articulate what is meant by an inversely proportional relationship and give examples to support their explanation.

2.1 Unknowns on one side

Small steps

- Solve one- and two-step equations and inequalities (R)
- Solve equations and inequalities with brackets (R)

Rationale

These review steps look back over learning from Books 1 and 2. Some students may only need a brief reminder before moving on to the more complex material provided in the chapter, whilst others may benefit from more practice. Students will need to be fluent with the equations and inequalities in this chapter in order to access the rest of this block effectively. You may wish to use the chapter as a vehicle for students to practise arithmetical skills, including fractions and directed numbers, or you could encourage the use of calculators if you want the students to focus on mastering the techniques. A wide variety of models are incorporated, and by now students may well have a preferred approach. Encourage those who are struggling with abstract representations to use their preferred model alongside symbolic recording whilst this is being embedded.

Things to look out for

- Trying to "spot" answers. Model examples where this is not easy to do and encourage students to use their mathematical knowledge to work towards solutions.
- Expecting (positive) integer solutions. Include all sorts of answers in your modelling examples.
- Treating inequalities in exactly the same way as equations. Encourage students to compare the questions with their answers, e.g. "What sign is in the question? What sign should be in the answer?"
- Errors when expanding brackets. Remind students to multiply all the terms inside the bracket by the term outside.
- Transcription errors. Encourage students to use substitution to check answers and, if there are errors, to work back through the calculation to identity where the mistake was made. This could be done in pairs.

Interleaving opportunities

- The techniques introduced in this chapter are built on throughout the remainder of this block.
- Equations and inequalities arise in many other areas of the curriculum, particularly the number work covered in Blocks 6 to 9.
- Equations with squared terms will be encountered in Block 11.
- Forming and solving equations and inequalities will be necessary in Blocks 13, 14, 16 and 17.

Models and representations

- Bar models are useful for helping students to visualise the steps needed to solve equations, and they can also be used for inequalities.
- Balances are a useful analogy, but are more cumbersome to draw as part of a solution. They are suitable for discussing the progress between steps of a solution for both equations and inequalities.
- Function machines are useful to support the "inverse operations" approach but, for equations of the form $a - bx = c$, the idea of self-inverse can be quite challenging.

Behind the questions

Are you ready?

The first question looks at algebraic notation and reminds students of the meanings of expressions with and without brackets. Questions 2 and 3 use function machines to remind students of working forwards and backwards, and the formal idea of an inverse is revised in question 4. Students then expand brackets in question 5 in readiness for the second part of the chapter. Any errors/misconceptions here will need to be addressed before the main exercises are tackled.

Practice 2.1A

1 provides fluency practice in solving simple one-step equations. Negative and non-integer solutions are included so that students are less likely to spot the answers and instead need to think about the process involved. Allow the use of calculators if appropriate. Students should check their answers by substitution, here and throughout the chapter.

2 moves on to inequalities. It is worth discussing with students what's the same and what's different about their approaches and answers to questions 1 and 2.

3 progresses to two-step equations of the form $a + bx = c$ (in various orders) and explores a variety of typical methods; this could be a useful discussion topic. Part c encourages students to look at the structure of the equations rather than solving them, and then check.

4 develops the previous question, now looking at equations of the form $\frac{x}{a} + b = c$. Part d is very important, and it is worth having a class discussion about how to choose the order of the steps in a solution.

5 moves on to solving inequalities and illustrates the similarity with solving equations.

6 provides fluency practice.

7 emphasises that inequalities do not have a single solution and explores the idea of a solution set.

8 encourages reasoning and the use of substitution to check solutions to equations.

9 ties up the exercise by providing mixed practice. This could be a good assessment question, either now or at a later date.

What do you think? 2.1A

1 Encourage students of think of equations of the form $x^2 = 16$ to disprove Faith's conjecture.

2 Number lines for showing solutions are explored here and revisited in Chapter 16.3.

3 looks at "enclosed" inequalities. You could look further at number lines if desired.

4 provides an opportunity to discuss the structure of equations in more depth.

Practice 2.1B

1 explores solving equations with brackets by expanding these or by division, and encourages students to look for factors. They should use the same reasoning for questions 2 and 3, which includes equations that need simplifying.

4 and **5** move on to inequalities.

What do you think? 2.1B

1 invites students to manipulate equations. They could be challenged to create their own, e.g. starting from $x = 5$, and then build up to increasingly complex multi-step equations.

2 looks at forming equations as a preview of Chapter 2.4.

Consolidate – do you need more?

This short section provides further fluency practice, if needed.

Stretch – can you deepen your learning?

1 looks at equations with more than one variable (explored in depth in Chapter 16.2).

2 Equations with unknowns on both sides will be covered in Chapter 2.3. This provides a good starting point for a discussion of students' ideas about how to approach such questions.

3 Students may need reminding that, for example, both 4 and −4 give 16 when squared.

Reflect

Students could create a display to consolidate their understanding and show their preferred approaches.

2.2 Inequalities with negative numbers

Small steps

- Solve inequalities with negative numbers

Rationale

This chapter covers inequalities with only a single step as the concept can be quite challenging. The idea of reversing the direction of the inequality sign is discussed, but the main approach is to add to both sides of the inequality to eliminate the term with a negative coefficient. This is introduced with equations first and then moves on to inequalities. This idea is also built on in the next chapter, which covers solving equations with unknowns on both sides. You may wish to use the model on page 53 with different related pairs of equations and inequalities.

Things to look out for	**Interleaving opportunities**
• Students sometimes find inequalities that do not start with the variable confusing, e.g. finding $5 > x$ more difficult to interpret than $x < 5$. Invest sufficient time so that students are able to read and represent positive inequalities correctly first. • The notion of reversing the direction of an inequality can be confusing as students may inadvertently apply it to expressions without negative coefficients. Model why this happens for negative coefficients by building up to it as illustrated in the text before the worked example in the Student Book. If appropriate, you can ignore the notion of reversing the direction entirely and just use rearrangement and adding to both sides of the inequality.	• Adding to both sides of an equation/inequality is developed further in the next chapter when solving equations with unknowns on both sides. • Negative coefficients also occur in Chapter 2.5.

Models and representations

- Algebra tiles are recommended to support the formal method of adding to both sides of an equation in order to eliminate a negative coefficient. Students should be familiar with the idea of a "zero pair" from their learning on directed number from Books 1 and 2. You may wish to remind students about zero pairs by using numbers before moving on to expressions. Online versions of algebra tiles are useful for demonstrating zero pairs.

Behind the questions

Are you ready?

1 and **2** check students' understanding of ordering directed numbers, as without this students may be unable to check solutions effectively.

3 and **4** check that students can fluently solve simple two-step equations and inequalities with both positive and negative solutions. Students need to be fluent with these skills to access the questions in this chapter.

Practice 2.2A

1 looks at three ways of solving an equation with a negative coefficient of *x*. Encourage discussion here, either in pairs initially and then as a whole class, or by using this as a worked example.

2 develops the ideas in question 1, with opportunities for students to gain practice. You may again wish to discuss the variety of approaches possible. Encourage the use of substitution to check answers.

3 Again, students should substitute values to check their answers. Discuss answers as a class and verify with other multiplications and divisions.

4 Here students solve one-step inequalities and observe the issues that arise with a negative coefficient.

5 continues the exploration. Students should notice that the −3 in part b does not affect the process of solving the inequality in the same way as the $-2y$ in part c.

6 provides practice with increasingly complex inequalities. Again, students should test a value in their solution set. Note that part e is an "ordinary" inequality with a negative solution; this is deliberate, to make students think rather than mindlessly apply a rule.

7 makes the link back to equations, this time increasing the complexity by including brackets.

What do you think? 2.2A

1 Students often make more errors when working with expressions such as $8 - (3x + 4)$ than with those such as $8 - 2(3x + 4)$. Encourage them to think that they are subtracting everything in the bracket, or to think of the former expression as $8 - 1(3x + 4)$ and to expand the bracket before simplifying.

2 provides further practice by looking at number puzzles to form the inequalities. Students need to take care with phrases such as "not more than", and realise that this means $\leqslant$ rather than $<$

Consolidate – do you need more?

1 challenges students to spot the common errors that are likely to arise.

2 and **3** provide more fluency practice and can be used now or when returning to revise the topic at a later date.

Stretch – can you deepen your learning?

1 introduces the idea that quadratic inequalities can have two distinct regions as the solution set. You may wish to include formal set notation, which was covered in the earlier books, so as to write the solutions as

$\{x : x \geqslant 5\} \cup \{x : x \leqslant -5\}$

2 looks at solutions to quadratic inequalities that can be written as a single set.

3 builds on the idea of using a number line to represent the solutions, as covered in the first "What do you think?" section in the previous chapter. Students could be challenged to convert their solutions from one form (e.g. words) to the others (set notation and number lines).

Reflect

Students reflect on their approaches to solving inequalities with negative coefficients.

2.3 Unknowns on both sides

Small steps

- Solve equations with unknowns on both sides
- Solve inequalities with unknowns on both sides

Rationale

Students who completed the Higher steps in Book 2 will already have met the ideas covered in the chapter, so this will be a reminder and extension for them. For those students who have not met the ideas before, this chapter makes no assumptions other than the prior learning in Chapters 2.1 and 2.2 of this book. The focus is on solving equations rather than forming them; forming equations is dealt with in the next chapter.

Things to look out for

- Difficulty in checking answers, which can be because students think of "=" as "makes" rather than "is equal to". Model substituting into both sides of the original equation to get two equal answers.
- Sign errors, with students "losing" negative signs between steps of complex solutions. Encourage them to check for this if their substitutions suggest that an answer is wrong.

Interleaving opportunities

- This builds on the learning from the two previous chapters.
- Further practice can be found in Chapter 2.4, which covers forming equations and inequalities.
- Equations and inequalities with unknowns on both sides will arise in many of the forthcoming chapters; they feature particularly in Block 9 as a result of angles being given in terms of algebraic expressions.

Models and representations

- Bar models are a powerful representation, as using two bars of equal length can enable students to visualise the next step of a solution by cutting the bars in a way analogous to "taking from both sides".
- Balances demonstrate the "taking from both sides" analogy even more clearly, although they are more cumbersome to draw.
- Cups and counters are a good way of modelling the solution to a class, and double-sided counters and/or inverted cups can be useful for modelling negative numbers and coefficients. They work best for integer solutions.
- As in the earlier chapters, all these models should be used alongside the symbolic notation that students should be encouraged to use, with the models supporting understanding rather than being used as an alternative.

Behind the questions

Are you ready?

1 and **2** check for the necessary fluency with the learning about equations from Chapters 2.1 and 2.2. Difficulty with negative coefficients will not preclude success in this chapter, but you will need to choose the questions carefully.

3 reminds students that the position of the unknown on either side of the equation does not alter the difficulty, as equations with unknowns on both sides can be reduced so that the unknown is on the right rather than the left, which is more commonly seen.

4 looks at inequalities of various types.

5 is a reminder of simplifying expressions with coefficients that involve directed numbers.

Practice 2.3A

1 and **2** model in detail the bar model approach to solving equations with unknowns on both sides by reducing them to a form with the unknown on one side only, as covered in Chapter 2.1.

3 looks at the form $ax + b = cx$, which students sometimes find difficult owing to lack of experience of equations of this form rather than inherent inability. Ask students to compare these with equations of the form $ax + b = c$ and $ax + b = cx + d$

4 and **5** discuss what to "take off" each side of an equation. Encourage students to generalise and conclude that taking off enough to reduce the equation to three terms whilst avoiding negative coefficients is preferable to always making the unknown appear on one particular side. Bar models can again be used to support.

6 and **7** bring in some negative terms. Students need to be careful not to "lose" or mis-copy signs, as illustrated in question 8, which also provides a good model of why checking is useful and how this works.

9 provides a mixture of examples for fluency practice. If more are needed, encourage students to write their own equations and challenge each other to solve them, as illustrated in question 1 in the "What do you think?" section.

What do you think? 2.3A

1 Students can generate further practice by choosing their own numbers as solutions and challenging each other. They could also include brackets here if you like.

2 links back to the previous chapter as students solve equations with unknowns on both sides where some of the coefficients are negative. This is a good starting point for a class discussion about when to add/subtract and how many to add/subtract.

3 uses number puzzles as a vehicle for forming equations with unknowns on both sides.

Practice 2.3B

1 extends students' learning by moving on to inequalities with unknowns on both sides, after a brief reminder of two-step inequalities.

2 and **3** encourage students to reflect on how many x to subtract from each side, and why.

4 and **5** mirror the approach of Practice 2.3A and introduce negative terms.

6 looks at a variety of types of inequality, adding in the idea of the smallest integer solution, which can easily be identified from the solution set. Students could be challenged to write inequalities given a smallest/greatest integer solution.

What do you think? 2.3B

1 In the same way as in the first "What do you think?" section, students now move on to inequalities with unknowns on both sides with negative coefficients. They may need reminding that a can be thought of as $1a$.

2 looks at the decisions needed when both coefficients are negative. This could be approached by multiplying by −1 and reversing the inequality, as hinted at by the error illustrated in part b.

Consolidate – do you need more?

This extra practice concentrates on equations, with some inequalities included. If this is used as revision/review at a later date, then question 1 provides bar model support.

Stretch – can you deepen your learning?

1 extends to looking at equations and inequalities involving fractions. Both suggested approaches need careful application as errors can occur with fraction arithmetic or from not multiplying all the terms.

2 and **3** look at incorporating brackets. Students may need support with $3 - (x + 4)$, for example by considering it as $3 - 1(x + 4)$.

4 looks at double inequalities. This could be a starting point for a deeper investigation if time allows, with students investigating when solutions are or are not possible.

5 The solution $x = 0$ is always an interesting discussion point.

Reflect

Students are encouraged to reflect on how this learning builds on the previous chapters and the on differences between equations and inequalities. This could form a useful display or revision notes.

2.4 Solving problems with equations and inequalities

Small steps

- Solve equations and inequalities in context
- Substitute into formulae and equations

Rationale

Students should now be fluent in the mechanics of solving equations and inequalities, so this chapter provides an opportunity for them to apply their skills in context. You may wish to use your knowledge of students who need extra challenge or support to bring in areas of the curriculum that need particular attention. The chapter focuses mainly on equations but there are some inequalities; the same principles are used to form them both. You may wish to challenge students to create their own questions, or adapt the existing ones using phrases such as "more than", "no less than", etc.

Things to look out for

- Students need to be sufficiently fluent at solving equations from the previous chapters.
- Students need to be fluent at forming expressions and to understand the meaning of algebraic terms. This is included in an "Are you ready?" activity and may be a useful whole-class activity prior to starting the chapter.
- Students may have forgotten the mathematical rules needed to form the necessary equations, and you may need to spend some time revising the key points.

Interleaving opportunities

- This links back to the techniques that students have been learning throughout this block.
- The use of formulae links to the next chapter.
- Students have the opportunity to revisit angles, probability and the use of formulae.
- This topic can be revisited in most future chapters of the book, e.g. Blocks 4, 9, 13 and 15. You can incorporate problems requiring equations into your examples in several other blocks too.

Models and representations

- Any of the models and representations from earlier in this block can be used to support the solution of the equations and inequalities that arise here.
- Algebra tiles are useful for forming expressions, particularly as the x and 1 tiles are already different and thus less prone to error when simplifying. The same is true for cups and counters, cubes, etc.
- Discuss with students that the algebraic representation is itself a model, i.e. a symbolic model as opposed to a concrete or pictorial one.

Behind the questions

Are you ready?

1 encourages students to remember angle relationships that will be needed later in the chapter.

2 looks at forming expressions from number puzzles in order to prepare for forming equations from them.

3 checks fluency with equations.

4 reminds students that the total probability of all possible outcomes of an event is 1.

Practice 2.4A

1 Students form and solve equations using the sum of the angles on a straight line; all the equations need some simplification of the numerical terms.

2 provides practice at forming and solving one-term equations and links to inverse operations. Students may increase the complexity with their own puzzles. Look out for puzzles involving "subtract from".

3 looks at possible errors when forming an equation. You could ask students to explain what the incorrect equations represent.

4 extends the topic slightly as students need to use their solution in the last part of the question.

5 leads to an inequality. Students may be tempted to use trial and improvement and may need to be encouraged to write an inequality; it is useful to compare the two methods. They may also need a hint to recognise the need for an integer solution.

6 Students may need reminding of the angle rules. Encourage correct mathematical vocabulary. Some parts require more than one value to be found, and these can be done in alphabetical order.

7 As in question 5, students may need a hint to recognise the need for an integer solution to the inequality.

8 The equation from this probability situation needs simplifying.

9 You can compare this with questions 5 and 7 and consider why non-integer solutions are possible here, noting that this time students are looking for a solution set.

What do you think? 2.4A

1 In part a, the use of x and $x + 3$ to form an equation with brackets may need modelling first. It might be worth discussing which solutions of the inequality in part b are realistic.

2 produces a series of equations and inequalities with unknowns on both sides.

3 interleaves both ratio and probability. You could discuss simplifying fractions when looking at part b, which can be extended to include other fractions. You can also revise vocabulary and notation, e.g. “P(prime) = …”

Practice 2.4B

1 to **3** look at substituting into different parts of formulae and expressions to form equations to solve. You can refer back to this when rearranging formulae in the next chapter. Question 3 is a rearrangement of Pythagoras' theorem, which will be covered in Block 11.

4 extends the topic as students need to choose the correct initial formula. They can check their choice by substitution.

5 encourages students to deal with the fraction in the formula for the area of a triangle by multiplying both sides of their equation by 2.

6 involves a more complex formula.

What do you think? 2.4B

1 Alert students to the link with factors.

2 looks back at gradients, providing a link with Block 1. Students' views on whether to use the formula or a sketch should be interesting.

Consolidate – do you need more?

This provides more practice at forming and solving equations. Note the bar model provided in question 4, which might otherwise be challenging.

Stretch – can you deepen your learning?

1 The first of these may need scaffolding, e.g. by suggesting that students call the smallest number x. You could compare assigning a letter to each of the three numbers and consider how that affects the solution.

2 Discuss why these are better approached as equations in one variable (with brackets) rather than two. Use this as a starting point for students to create similar number puzzles, or find some in books/online.

3 There is no need to link this to solving quadratics, although you could do so if appropriate. The question provides good practice in directed number calculations. Note that part b can be solved without algebra, as the total age must increase by 2.

Reflect

This reflection is designed to consolidate students' understanding of vocabulary and different mathematical constructs. You could include words such as “expression”, “term” and “variable”, if appropriate.

2.5 Rearranging formulae

Small steps

- Rearrange formulae (one-step)
- Rearrange formulae (two-step)
- Rearrange complex formulae including brackets and squares (H)

Rationale

Students sometimes do not make the connection between solving equations and rearranging formulae, so this chapter makes the link explicit. A variety of known formulae are used before moving on to more abstract constructs that behave in the same way. Students may find the language difficult, so interchange the use of "rearrange", "express in terms of" and "change the subject of" to familiarise them with these phrases.

Things to look out for

- Students may need assistance to understand the order in which to perform operations to make another letter the subject. Link back to equations and ask them how they would solve the equation if the current subject was, say, 20.
- Students may need support with fractional notation, for example in understanding that $r = \frac{1}{2}d$ and $r = \frac{d}{2}$, say, are equivalent. Use this in the familiar context of the radius being half of the diameter of a circle.
- The convention of starting with the subject should be discussed; see the worked examples in the Student Book.

Interleaving opportunities

- This builds on the earlier work in this block.
- Students can revisit the equation of a straight line and rearrange more complex forms in order to find the gradient and intercept.
- You can return to this topic whenever a new formula is encountered, e.g. the volume formulae in Block 4 and Pythagoras' theorem in Block 11.

Models and representations

- As students will be familiar with using bar models to help with the solution of equations, they can also look at their use to represent formulae. Start with one-step formulae, where $t = 5p$ and $p = \frac{t}{5}$, for example, can easily be seen to be equivalent. Once the ideas become clear, students can move on to using just abstract representations for more complex formulae.

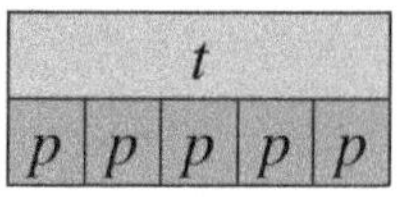

- Function machines are useful models, and they provide a clear link with inverse operations, although they are more problematic when the new subject has a negative coefficient in the original formula. This is addressed in the first "What do you think?" section.

Behind the questions

Are you ready?

1 checks the understanding of inverse operations necessary for this topic.

2 looks at using a formula to find each of the three letters; you could refer back to this when students have learnt how to change the subject, and compare solutions and approaches.

3 is structured to make students think about the first step in solving the equations, a thought process they will need for rearrangement.

4 reminds students of the inverse nature of squaring and finding the square root.

Practice 2.5A

1 It is crucial that students understand the meaning of the word "subject" and can identify the current subject of a formula.

2 looks at three methods in the familiar context of the perimeter of a square. Comparing these is a good starting point for a class discussion.

3 Note that, although the formulae are presented as bar models, students can use any approach to find the solutions.

4 moves on to two-step rearrangement, again providing a variety of models. The gradual variation in the formulae students are asked to rearrange forms another good discussion point.

5 Students may need reminding that π is a constant not a variable.

6 looks at a wholly abstract formula expressed in terms of letters. Emphasise that the process remains the same.

7 Students can find this type of question challenging. Link back to question 5 by comparing $C = \pi d$ and $C = 2\pi r$.

What do you think? 2.5A

1 links back to solving equations and inequalities with negative coefficients in Chapter 2.2. Again, this could be a good question to discuss with the class, in particular noting how the bar model in part c can work with negative numbers. The formulae that students need to rearrange in part d have slight variations to encourage them to consider what's the same and what's different.

2 Students may need support with the notion of a self-inverse. This can be demonstrated numerically, e.g. with 10 – 4 = 6 and 10 – 6 = 4. Compare with the other methods and discuss which students prefer.

3 links to finding the gradient and intercept when the equation of a straight line is not given in the form $y = mx + c$. The fact that it is called the "equation of the line" when it is also a formula connecting y and x provides another interesting talking point.

4 links to using substitution for checking solutions to equations.

Practice 2.5B

1 looks at using function machines and symbols to rearrange formulae with squared terms. Understanding the difference between $2y^2$ and $(2y)^2$ is crucial for other formulae later, so this is addressed explicitly here. Students then practise this in question 2.

3 makes the point that different formulae can be equivalent. Question 4 then looks at variations of formulae that might result in different, but equally valid, answers.

5 develops this with increasingly complex formulae. Careful consideration is needed about the order of inverse operations to make x the subject.

6 combines the ideas of question 1 with the variation of questions 4 and 5 to look at complex formulae with square roots, and with squares and square roots.

What do you think? 2.5B

1 looks at a simple rearrangement of a multiplicative formula in preparation for question 2.

2 Students need to make the denominator of a fractional formula the subject. Link this to solving equations such as $30 = \frac{90}{x}$ and to rearranging, say, $a = 10 - bx$.

3 looks at fractions and squaring.

4 is more complex, but the hint should help. A second hint would be to find a common denominator.

Consolidate – do you need more?

Further practice is provided. Question 6 again links to rearranging the equation of a straight line.

Stretch – can you deepen your learning?

1 looks at a formula in words and provides practice at solving an equation with unknowns on both sides.

2 Students form an equation from a problem expressed in words and substitute one formula into another.

3 extends to formulae where the subject appears twice, but no factorisation is needed.

Reflect

Students reflect on the order of the steps when rearranging a formula and on the links with solving equations.

3.1 Properties of number

Small steps

- Explore factors, multiples and primes (R)

Rationale

This short chapter provides a review of prior learning about factors, multiples and primes. Although these terms are important in their own right, students will need to rely on this knowledge when they explore conjectures about numbers in the other chapters in this block. Prime decomposition is revisited, and both the "ladder method" and the "factor tree" method, which was used in the earlier books, are demonstrated. By the end of this chapter, students should be fluent at recognising prime numbers, even numbers (multiples of 2) and odd numbers (non-multiples of 2) and at identifying other multiples and factors.

Things to look out for

- Confusing the terms "factor" and "multiple". Emphasise that a multiple is the result of a multiplication.
- Thinking that 1 is a prime number. Emphasise the definition of "prime" as having exactly two factors rather than "cannot be split up".

Interleaving opportunities

- Multiplication and division skills
- Using arrays and bar models to solve problems
- Areas of rectangles
- The terms factor and multiple will be met again when HCF and LCM are revised in Block 6

Models and representations

- Arrays are very useful for showing the factors of a number and to build up multiples. An array like this can be used to generate as many sentences as possible using the words "factor" and "multiple" to check students' understanding. You can also explore the arrays formed by a prime number.

- The multiplications and divisions shown by a bar model can also be used to generate sentences using multiples and factors. You can also work the other way round, by asking students to "Draw a bar model (array) to show that __ is a factor/multiple of __".
- Factor trees are the main vehicle used for finding prime decomposition, which builds on the learning from earlier books. As a comparison, the ladder method is included, and it might be useful to compare and contrast the methods in a class discussion. The systematic approach of working through the primes in order that the ladder method employs may be useful for students who find it difficult to spot factors.

Behind the questions

Are you ready?

Question 1 is a very simple question to promote discussion of multiples and patterns within them. Question 2 looks at the crucial need to recognise even (and therefore odd) numbers, and question 3 provides some multiplication fluency practice. If students are not confident with the times tables, they could access the content of the chapter using a calculator so that they still appreciate the learning points.

Practice 3.1A

1 Students can draw the arrays for support, if needed. Encourage them to make sentences using the words "factor" and "multiple", as well as to write down the multiplication and division facts.

2 Again, students can be challenged to use the word "multiple" as well as the word "factor". You could also represent the arrays as rectangles of small squares and make the link to area.

3 Students should be able to answer this question using their knowledge of times tables. You could use this as the basis for a discussion about how to recognise multiples of numbers and include, for example, using repeated division by 2 to identify a multiple of 8.

4 and **5** develop the ideas of the chapter and go on to revise the concept of a prime number, which can be illustrated using an array with a single row or column only.

6 revises the use of factor trees to find the prime decomposition of a number, as covered in Book 1. Look out for students including unnecessary 1s, and splitting by addition rather than multiplication.

7 looks at less common factorisations of numbers that have exactly two prime factors (called semi-prime). Students could investigate numbers of this form further, if desired; for example, 33, 34, 35 is a set of three consecutive semi-primes.

8 explores using one prime factorisation to find that of a related number. Again, this could be extended.

9 introduces the ladder method, dividing by each prime in turn as many times as possible before moving on to the next prime until the decomposition is complete. The advantage of this method is that, if applied correctly, the prime factors are in order and it easily leads to the form $a^m \times b^n \times \ldots$ for primes a, b … .

10 explores the ideas from question 8 algebraically.

What do you think? 3.1A

1 challenges students' understanding of the link between the words "factor" and "multiple". Students could use numerical and/or algebraic examples.

2 extends the idea of question 8 in Practice 3.1A by using more complex numbers. Students could create their own similar questions and challenge each other.

3 provides a link to algebraic factorisation. This is a good opportunity to revise this, if necessary.

4 explores the distribution of primes and recommends an internet search to deal with the increasing numbers.

Consolidate – do you need more?

Here more fluency practice is provided. Question 1 explicitly looks at the area model as an alternative to the array.

Stretch – can you deepen your learning?

1 Encourage students to explore odd and even numbers, primes, etc.

2 This prime generator (for $n = 1$ to 39) should provoke interesting discussion.

3 looks ahead to generalising the number of prime factors of a number given its prime factorisation.

4 Encourage students to explore the divisibility of exponents to arrive at the generalisation.

Reflect

You could discuss students' preferred methods for factorising large numbers. Is it the same or different from how they would tackle smaller numbers?

3.2 Looking for truth

Small steps

- Explore whether a statement is true or false
- Explore whether a statement is always, sometimes or never true

Rationale

As an introduction to exploring conjectures, this chapter starts by looking at statements that are either true or false and that students can identify using either their existing knowledge or some exploration/testing. The length of the chain of reasoning they need to employ increases gradually throughout the block, with "always true, sometimes true, never true" extending to three choices and needing more consideration. Students could create and test their own conjectures in this format, to develop their thinking and reasoning skills further.

Things to look out for

- Only using one example to "prove" that something is true. Practice 3.2B, in particular, demonstrates why a single example is not sufficient.
- Not understanding the term "counterexample". Model using counterexamples, and encourage students to see how it is much easier to disprove a statement (only one counterexample is needed) than to prove it (proof by exhaustion is often impractical, so algebraic or other methods are needed).
- Difficulty in accessing the questions that revisit concepts from some time ago. These questions have been designed to look at familiar concepts, but some reminders might be needed.

Interleaving opportunities

- Many of the conjectures rely on the factors, multiples and primes material covered in Chapter 3.1.
- Arithmetic skills and the equivalence of fractions, decimals, percentages are included.
- Equations and simple inequalities, equations of straight lines, probability and shape/angle facts are all used as vehicles to develop students' reasoning.
- You could also use this chapter to revisit any topic that you would like your students to practise, e.g. area/perimeter/ratio.

Models and representations

- You can also use the same models as in Chapter 3.1, for example arrays and area models, to look at the conjectures involving factors, multiples and primes.
- The structure of odd and even numbers becomes increasingly important as this block develops, so using a "two rows" model such as this to represent an even number and the same image with one extra added to one of the rows to represent an odd number will help students to link these to the later algebraic forms of $2n$ and $2n + 1$.

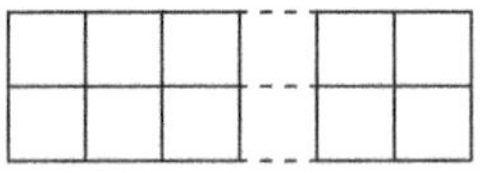

Behind the questions

Are you ready?

Questions 1 and 2 check competence with the language of the previous chapter. Question 3 revisits the equation of a straight line from Block 1, so a quick reminder may be needed. Question 4 reminds students about probability, which features in a small number of questions in the chapter. In particular, students need to know that probabilities must be between 0 and 1, and that events are not always equally likely.

Practice 3.2A

1 Students could answer these by performing the calculations or by observing their structure. In part a, for example, 80 + 20 = 100, so (greater than 80) + (greater than 20) must be greater than, not equal to, 100.

2 Students may need reminding about fraction, decimal and percentage equivalence. Encourage alternative approaches to calculation, e.g. observing that $\frac{3}{5}$ cannot be 35% because $\frac{3}{5} > \frac{1}{2}$ but 35% $< \frac{1}{2}$ (50%).

3 develops the idea of using reasoning rather than calculation to determine whether the statements are true. Comparing with $\frac{1}{2}$ is a particularly useful strategy for parts b i and b iii. All provide good discussion points.

4 requires students to reason about the language used and the approach to finding the solution, e.g. part a must be answered by looking at the complete lists rather using than a quicker way.

5 Students may need a reminder about drawing sample spaces, especially for combined events such as flipping a coin twice. This will be studied in detail in Block 15.

6 91 = 7 × 13 but is often mistakenly thought of as being prime.

7 Students should check by substitution rather than by solving (especially as there is a quadratic!).

8 looks at a common error when finding the height of a triangle given its area.

9 Students may need a reminder of their learning from Block 1.

What do you think? 3.2A

1 By working systematically, students should come up with the square numbers.

2 Encourage students to test conjectures of this form with positive and negative integers, numbers between 0 and 1, 0 itself, etc.

Practice 3.2B

1 Working systematically should generate sufficient examples for students to appreciate that statements are not always either true or false for all values.

2 develops the ideas from question 1, but students need to make the choice themselves.

3 uses a sequence to develop students' reasoning, again using the vocabulary of Chapter 3.1.

4 revisits shape work from Books 1 and 2. Encourage students to draw sketches.

5 uses directed number arithmetic, which may need a brief reminder.

6 Note that, at this stage, descriptions rather than formal proofs or illustrations will be expected.

What do you think? 3.2B

1 looks at using more complex algebraic language, but tests the same reasoning. Encourage students to use substitution.

2 provides an opportunity to revisit the language of averages and the range. Again, encourage systematic working and looking at extremes: "What happens if you change one number by 1? … by 100?"

3 Part a is a reminder about units and part b should again be approached systematically.

4 Students could explore a variety of shapes, and you could use these questions to keep their knowledge of formulae sharp.

Consolidate – do you need more?

These extra questions use the same ideas and areas of mathematics as the main exercises.

Stretch – can you deepen your learning?

1 progresses to slightly more formal and algebraic problems.

2 looks at generalisation using the equation of a straight line. You could compare the answers with equations given in other forms, for example $ax + by = c$.

3 builds on Chapter 3.1 and can be extended and generalised to the products of more prime factors.

Reflect

Students should include a reference to a systematic approach, and looking at extreme or boundary values, such as 0 and 1.

3.3 The journey to proof (1)

Small steps

- Show that a statement is true
- Conjectures about number

Rationale

In this chapter, students extend their learning to showing more formally whether statements are true or false, including using manipulatives to illustrate results. They will continue to justify whether statements are true or false, or are always/sometimes/never true. This will be developed into using algebraic proof in the next chapter. As before, students should be encouraged to think of their own conjectures and to look at alternative ways of showing their results.

Things to look out for

- Always doing calculations when results can be shown more easily by observing the structure of the problem. Model alternatives and encourage students to look for short cuts before starting calculations.
- Sign errors, or other errors with algebraic manipulation. Use algebra tiles to support understanding.
- Trying to start with the result in a "show that" problem, instead of working towards the result from the given information. This needs careful modelling.

Interleaving opportunities

- Factors and multiples, especially odd and even numbers
- Fraction, decimal and percentage equivalence, and calculations involving fractions, decimals and percentages
- Forming and interpreting expressions
- Expanding brackets and factorising
- Again, you could use this section to revisit any topic that you would like your students to practise, e.g. probability/area/perimeter/ratio.

Models and representations

- All the models that students are familiar with from the earlier books, such as bar models, cubes and counters, are used to represent algebraic constructs. In particular, two equal rows of a bar model are used to show an even number (showing that it can be split into two equal parts), and then an odd number is represented as being one more (or one less) than an even number.
- Algebra tiles are also useful for illustrating expanding brackets, collecting like terms, etc. You can use the "ones" algebra tiles (or any other double-sided counters) to model directed number arithmetic.

Behind the questions

Are you ready?

Question 1 looks at odd and even numbers in preparation for the "two rows", "two rows plus one" model later in the chapter. Questions 2 and 3 look at expanding brackets and factorising, which students will need for some simple algebraic deduction. Question 4 checks proficiency with adding/subtracting directed numbers, which will be needed in simplifications later.

Practice 3.3A

1 This opener is useful for discussing the pros and cons of showing a result by calculation and other methods. Encourage students to look for factors and to observe that both calculations are equivalent to $8 \times 4 \times 10$. You may wish to remind students about commutativity and associativity.

2 Again, factorisation is the easiest approach here but, for students who struggle with this, showing that the calculations are equal because they have the same result is perfectly valid.

3 You may need to remind students that decimals are easily converted to tenths, hundredths and thousandths.

4 You can have a discussion that compares the various methods, e.g. using equivalent fractions, decimals or percentages, diagrams, distances from 1, the fact that $\frac{1}{6} < \frac{1}{4}$, etc.

5 For the first part, calculation is probably the best approach, whilst for the others it is useful to observe that they reduce to the same overall calculation without working out the actual answers. The fact that *a*% of *b* is equal to *b*% of *a* is proved in Chapter 3.4, but students may make the observation (or even prove it) here.

6 You can discuss how the result can be reached in multiple ways, e.g. by adding all four sides or from 2 × length + 2 × width, or 2 × (length + width). This is a useful lead-in to the next question, involving brackets.

7 You may wish to model some questions like these first. Some degree of algebraic fluency is required, and you can use similar questions with altered levels of difficulty as appropriate for your class. Note that although parts a and b can be answered by working from left to right, part c could be shown by expanding and simplifying both expressions and observing that the results are identical, which avoids the need to factorise.

8 The expected approach is by substitution.

9 The numerical results are not needed in parts a and b, but the expressions are required in part c.

What do you think? 3.3A

1 involves simple algebraic manipulation.

2 and **3** build, once again, on the skills of Chapter 3.1.

4 makes a good link with the previous year's work on percentages and fractions.

Practice 3.3B

1 students should now be familiar with the notion of a counterexample.

2 provides a starting point for discussion. The focus is on understanding the structure to prepare for the algebraic form in Chapter 3.4.

3 builds on question 2 and can be extended to allow students to form and test their own conjectures.

4 reminds students that an example is not a proof and then develops the representation to look at multiples of 3, using 3 rows in the bar models, etc.

5 extends to multiplication. Students may need support to see that 4 objects must be 2 × 2 objects. You could extend this to odd numbers too. These is covered algebraically in question 9 of Practice 3.4B.

6 makes some conjectures about directed numbers. Counterexamples may be useful here.

What do you think? 3.3B

1 takes the conjectures about odd and even numbers further by looking at larger sets of numbers. Encourage students to generalise and form conjectures for themselves.

2 looks at the same ideas as question 1 in another context.

Consolidate – do you need more?

This provides extra practice in the same style as the main exercises.

Stretch – can you deepen your learning?

1 links the learning of this chapter to sequences and types of number.

2 Students may need guidance to think of the number as $100a + 10b + c$, which leads to a fairly simple proof.

3 and **4** develop the idea further by looking at the converse and the fact the result also holds for multiples of 9 and 3. You may prefer to come back to these questions after completing the more formal examination of algebraic proof in Chapter 3.4.

Reflect

Students reflect on the different possible approaches. Encourage them to use diagrams or manipulatives to illustrate their answers.

3.4 The journey to proof (2)

Small steps

- Expand a pair of binomials
- Explore conjectures with algebra

Rationale

Students may have expanded two binomials as a Higher step last year. If so, then only brief revision of the first exercise is needed before moving to the more formal proof with algebra in the second part of the chapter. The skill of expanding binomials is needed both in Practice 3.4A here and for the conjectures on 100 squares in Chapter 3.5. Links are made with multiplying numbers to remind students that algebra is generalised numbers. Students then move on to developing formal algebraic proof, which is linked closely to the types of conjectures illustrated diagrammatically in the previous chapter.

Things to look out for	Interleaving opportunities
• Errors in multiplication. Students should not move on to purely abstract representations until they are ready, and supporting models such as the area model are very helpful for building up understanding. • Errors in simplification, particularly with directed numbers. Use tiles/counters for support, as necessary.	• Expanding a single bracket • Expanding two binomials is needed in Chapter 3.5.

Models and representations

- Both algebra tiles and the related area model are very useful for illustrating the expansion of two binomials. These models should be used alongside the symbolic representation so that students make the link between them.
- As in the previous chapter, cubes and counters, etc. are good for showing multiples of 2, 3, etc.

Behind the questions

Are you ready?

In question 1, students need to be fluent with the expansion of a single bracket, both with numerical and with symbolic multipliers. Questions 2 and 3 check competence at identifying and collecting like terms. Question 4 provides practice at expanding two sets of brackets whilst question 5 looks at factorising in preparation for showing, for example, that $9k + 6m$ is a multiple of 3.

Practice 3.4A

1 models the use of an area model to find the expansion of a pair of binomials. Different letters are used in each binomial to initially avoid the added complication of needing to simplify. The link is made with multiplication of numbers; you may wish to refer to the grid method of multiplication if your class are familiar with it.

2 provides fluency practice in the same style as question1. Students may quickly become less reliant on the diagrams.

3 moves on to using the same letter in both binomials, with the aid of both the area model and algebra tiles. It is suggested that you find out which your students prefer. You could use these models alongside other more abstract representations. Students will now need to simplify their answers.

4 provides fluency practice. The questions involve positive terms only.

5 shows that multiplication is commutative even when the numbers are represented as expressions.

6 challenges a common misconception about squaring, and again links back to numbers.

7 Negative terms are introduced through algebra tiles.

8 challenges a common misconception.

9 provides some fluency practice in all the types of expansion met so far.

What do you think? 3.4A

1 challenges students to look at expansions with more terms if they are confident and this is appropriate.

2 explores the difference of two squares and how this can be used to simplify calculations. There is no need to learn about this in detail at this stage; it is here as an observation to spark interest and show a purpose.

3 advances to more complex expressions. Again, only use if appropriate as this will be developed further at KS4.

Practice 3.4B

1 sets the groundwork for algebraic proof involving odd/even/consecutive numbers by ensuring that students understand and can interpret expressions. It is recommended that these are discussed alongside the corresponding concrete representations used in Chapter 3.3.

2 takes the ideas of question 1 further by asking students to interpret the given expressions.

3 provides practice at forming expressions, again to underpin that, for example, $4n$ means 4 multiplied by n, so if n is an integer, then the expression is a multiple of 4.

4 further develops understanding of algebraic expressions showing multiples. In particular, it is important that, for example, $6k$ is seen as a multiple of 2 and 3 as well as a multiple of 6, and this can be combined with the results from question 1 that, for example, $10k - 2$ must be even.

5 looks at the formal proof that odd + odd = even and emphasises that the two odd numbers may be different. This is a good point for class discussion before moving on.

6 builds on the earlier questions. Now students can tackle these with less scaffolding.

7 and **8** look at slightly more complex conjectures using algebraic notation.

9 moves the conjectures on to those about products, some of which need the expansion of two binomials.

10 invites students to form their own conjectures. You may wish to direct these.

What do you think? 3.4B

1 looks at squaring.

2 Students may need to be reminded to use different letters and to factorise.

3 formally proves the results that were shown with objects in the previous chapter.

Consolidate – do you need more?

This provides further practice related to both exercises in the chapter.

Stretch – can you deepen your learning?

1 Students perform more complex expansions of trinomials and three binomials. It may be worth discussing how to set out their solutions. A lot of care is needed!

2 Students generalise a result with very little scaffolding.

3 investigates the famous unproved Goldbach conjecture.

Reflect

Students reflect on the key points of the chapter. They should be encouraged to use algebraic and other representations in their answers.

3.5 Searching for pattern

Small steps

- Explore the 100 square
- Expand a pair of binomials
- Explore conjectures with algebra

Rationale

In this chapter, students explore patterns within a hundred square and, if appropriate, other square and/or rectangular grids of different sizes. These are closely connected to the GCSE investigations "T-totals" and "Opposite corners". Students start by looking at blocks in different positions on the grids and work out numerical totals of the numbers in the blocks. You could encourage them to look at the totals of consecutive blocks on the grid before moving on to making algebraic generalisations, noting, for example, that the total for a 5-square block increases by 5 and considering the connection between the expression for the total and the rule for a linear sequence. These activities can be extended to look at the effects of rotations and reflections of the blocks. The questions in the exercises are initially structured and then the scaffolding is gradually removed. You can add or remove support as appropriate for your class. There are relatively few questions in this chapter, but many of them are open-ended and could be explored over several lessons if required.

Things to look out for

- Errors in finding the total of the algebraic expressions – in particular, remind students to include the original number.
- Errors in subtracting pairs of binomials of the form $(x + a)(x + b) - x(x + c)$. Encourage students to look carefully at the signs, or you may wish to limit some students to the investigations involving only addition.
- Students may need support in understanding the suffix notation used throughout this chapter.

Interleaving opportunities

- Collecting like terms
- Expanding binomials
- Factorising into a single bracket
- Link back to Chapter 3.4 to remind students how to write expressions representing odd or even numbers, multiples of 3, etc.
- Link totals of consecutive blocks to terms in a linear sequence.

Models and representations

- The key representation in the chapter is the hundred square, and it may be useful to provide students with multiple blank copies (on a single sheet) that they can use for their investigations.
- If students move on to look at different sizes of grids, these could also be provided, or students could create their own on spreadsheets.

Behind the questions

Are you ready?

Question 1 checks the forming of comparative expressions so students can later form generalisations in the grids. Question 2 checks expansion for the "opposite corner" type investigations. Question 3 looks at simplification of expressions with many terms, as will be required for finding general terms.

Practice 3.5A

1 This highly structured question looks at blocks on one row, so only terms x, $x + 1$, $x + 2$ are included. You may wish to also look at consecutive blocks, such as R_{14}, R_{15}, R_{16}, etc.

2 is a slightly more complex example retaining much of the structure. Students may need to be reminded to factorise the answer to part f. They may also need a hint to form and solve an equation for part g. You could extend this by looking at other given totals, so that students appreciate the need for integer solutions, or by forming and solving inequalities to find the first/last block over/under a given total.

3 develops these ideas with less structure so students need to work more independently.

4 moves to a different type of investigation which looks at the difference between products of numbers and then expressions. Students need to be confident at expanding binomials and subtracting multiple terms.

What do you think? 3.5A

1 is an open-ended development of question 4 of Practice 3.5A. You might wish to add more structure or ask students to work in groups to look at squares, $m \times 2$ for varying m, or $2 \times n$ for varying n, etc., and then report their findings to the class.

2 is also open-ended to encourage students to work systematically.

Consolidate – do you need more?

1 This adapts question 1 of Practice 3.5A by looking at vertical rather than horizontal blocks. The highly structured nature will support students who find the abstract nature of this topic difficult, or it may be useful if you are returning to this activity some time after completing the original exercise.

2 repeats question 2 of Practice 3.5A with a different shape.

Stretch – can you deepen your learning?

1 invites students to investigate the same shape on different sizes of grids, with very little structure. Students should work systematically and look for connections between the different grids to find general rules. They could also predict and test rules for the shapes they worked on earlier in the lesson but now using different sized grids.

2 generalises the earlier “opposite corners” investigation.

Reflect

These questions encourage students to think about the strategy rather than the process of working through investigative tasks.

4.1 Into three dimensions

Small steps

- Know names of 2-D and 3-D shapes
- Recognise prisms (including language of edges/vertices)

Rationale

The purpose of this chapter is two-fold. In the first section, there is the opportunity to revise names and key properties of 2-D shapes, including symmetry, angle facts and diagonal properties. Throughout, students should be reminded of the need for correct language when describing properties of shapes. Also in the first section, students will meet 3-D shapes for the first time in their secondary studies. The section focuses on naming shapes and on identifying and counting edges, vertices and faces.

In the second section of the chapter, students will explore properties of 3-D shapes in a little more detail. In particular, they will explore which shapes are prisms and which are not. At this stage, 3-D objects that they can hold and manipulate will be helpful. Edges, faces and vertices will be revisited, giving students a chance to explore the relationships between them.

Things to look out for

- This chapter is another chance to address misconceptions about shapes, for example the number of lines of symmetry in a rectangle.
- The vocabulary of 2-D and 3-D shapes can be confusing (for example, many students will describe an edge as a side because it links more directly to 2-D descriptions), so it is important to reinforce the correct terminology.
- A cylinder is often viewed as a prism but, although it has many of the same properties, it does not have the rectangular faces that characterise a prism.

Interleaving opportunities

- There are plenty of opportunities to explore and revise properties of 2-D shapes, including symmetry, properties of quadrilaterals and triangles and the language of shape in general.
- Algebraic notation is utilised later in the chapter, when generalising the relationship between edges, faces and vertices.

Models and representations

- Geoboards can be used to emphasise key properties of 2-D shapes.
- 3-D shapes made of wood, plastic or foam are readily available and will support students who struggle with 3-D visualisation.

Behind the questions

Are you ready?

These questions provide an opportunity to check students' current knowledge of shape names and properties. The focus is on 2-D shapes, but question 4 is a chance to see how much students recall about 3-D shapes.

Practice 4.1A

1 to **4** provide ample opportunities to revise the properties and vocabulary of shapes. There is plenty of material to encourage discussions in questions 2 and 4. Throughout, sketching shapes or using a geoboard can be useful.

5 is an excellent starting point for a discussion – there are many properties that can be discussed, including angles, side lengths, symmetry, diagonals and parallel lines. The diagrams are deliberately bare to encourage students to come up with the properties of each shape.

6 is an interleaving opportunity to recall properties of angles in quadrilaterals and angles in parallel lines.

7 to **9** deal with the basic properties of 3-D shapes, including counting faces, edges and vertices; looking at relationships between them will come later.

What do you think? 4.1A

1 is an excellent discussion task that encourages students to be extremely precise about their use of vocabulary and the properties that define each shape. Further questions can be asked, such as "Why is a square a rectangle but a rectangle is not a square?"

2 provides a little interleaving task involving Venn diagrams and a chance to explore a property of quadrilaterals that is often forgotten – their diagonals. Geoboards would be very useful here.

3 allows students to develop their reasoning skills using the newly learnt concepts of faces, edges and vertices.

Practice 4.1B

1 is a quick task to revise and recall names of 3-D shapes.

2 reinforces the concept and definition of a prism as having a constant cross-section, which need not be a regular or familiar shape. It's important that students recognise other prisms beyond those with triangular and other polygonal faces.

3 builds carefully towards recognising the relationship between edges, faces and vertices. Students are encouraged to apply the relationship to other shapes to see that it doesn't just work for prisms. Part d provides a chance to practise algebra skills.

4 and **5** provide opportunities to use the newly discovered relationship and apply it to reasoning-style problems.

6 reinforces that rectangular faces are a key property of prisms and can be used to establish that the number of rectangles will always be the same as the number of sides of the shape forming the cross-sectional face.

What do you think? 4.1B

1 is a chance to address a common misconception about prisms; it is important to emphasise that the cross-sections are not just the same shape but are congruent.

2 reinforces that cuboids are simply rectangular prisms and hold all the same properties as other prisms.

3 is an excellent discussion task: students need to recognise that a cylinder is not a prism, by definition, but has many of the same properties as a prism (which will become apparent when looking at volumes in Chapter 4.5).

4 allows more confident learners to extend themselves by generalising relationships using algebraic notation.

Consolidate – do you need more?

This section provides further questions and opportunities to consolidate the language and properties of 2-D and 3-D shapes.

Stretch – can you deepen your learning?

1 is a chance to consider shapes that don't conform to the $F + V = E + 2$ relationship. Cylinders and spheres have no vertices and, in fact, no edges.

2 extends question 1 by considering more deeply the definitions of faces, edges and vertices: an edge is where two faces meet and a face is defined to be a flat (not curved) surface. Thus a cone (and also a cylinder) has no edges.

3 encourages students to sketch in three dimensions and to use properties of pyramids and prisms.

4 is a chance to extend students by using algebraic notation for generalising properties of pyramids.

5 offers students a chance to investigate the Platonic solids (perhaps for a homework).

Reflect

1 and **2** both encourage students to reflect on the key take-aways of the chapter – what defines a prism and the relationship between the numbers of edges, faces and vertices in a 3-D shape.

4.2 Nets and other diagrams

Small steps
• Construct accurate nets of cuboids and other 3-D shapes • Sketch and recognise nets of cuboids and other 3-D shapes • Explore plans and elevations

Rationale

In this chapter, students will explore different ways of representing 3-D shapes in two dimensions. The focus will be on nets (both accurate diagrams and sketches) and also on plans and elevations. Students will be expected to move between 2-D and 3-D representations, so there is also the opportunity to practise isometric drawings.

When drawing nets, students should be encouraged to see that there isn't just one way of drawing a net for a given shape and realise that there are many possible configurations. When learning to sketch, students need to appreciate that although the diagrams do not need to be drawn accurately, the measurements should be correctly marked on them.

When exploring plans and elevations, you could show students some real-life applications such as architectural drawings and discuss the advantages of this method of representation.

Things to look out for	**Interleaving opportunities**
• Some students will struggle enormously with these representations. They may benefit from using actual 3-D objects and linking cubes to help them to explore plans and elevations. • It is worth reiterating to students that the side and front elevations are interchangeable if no clear direction has been given. • Most students are able to draw nets of cubes and cuboids but find prisms and other shapes challenging. Students should be allowed to make their own nets to test their conjectures about lengths of sides, etc.	• When drawing accurate nets, students will need to recall properties of shapes (such as isosceles triangles). • Once they have covered constructions in Block 5, there will be opportunities to review accurate constructions of nets using a ruler and a pair of compasses.

Models and representations

- Everyday objects such as cereal boxes, tins and their labels, and other packaging should be used to give students a chance to explore nets in a hands-on manner.
- Linking cubes will support work on isometric drawings and plans and elevations by enabling students to make the shapes they are being questioned about.
- Folding geometry kits are available from educational suppliers which allow students to remove a net from a 3-D shape and compare the lengths and faces with the actual 3-D shape.

Behind the questions

Are you ready?

1 and **2** review the vocabulary of faces and the properties of cuboids and prisms.

3 and **4** are warm-up questions that provide some straightforward examples of accurate drawing.

Practice 4.2A

1 and **2** provide practice at drawing accurate nets from 3-D drawings, whilst question 3 encourages students to think about this in reverse. Throughout, students should be reminded that there is more than one possibility for a net of a given shape, and they could be encouraged to share different ones.

4 helps students to develop their spatial reasoning. Some will find these questions challenging and may need to work with a net. However, they should be encouraged to conjecture first before using the nets.

5 is another chance to draw an accurate net, this time with triangles, which may prove more challenging.

6 to **8** explore nets of cuboids in a little more depth to allow students to further develop their reasoning skills. In question 6, students could be invited to consider how they could correct the nets which don't work.

9 is a chance to recap the names of 3-D shapes, but this time from their nets only.

What do you think? 4.2A

1 is a challenging investigation. It's important to define the meaning of "different" to clarify that different orientations of the same shapes don't count. There are 11 different nets, but students do not need to be told this.

2 involves a more challenging accurate drawing in which students will need to use their spatial reasoning skills.

3 allows some learners to explore the net of a cylinder before it is met formally in Chapter 4.4.

4 explores the properties of a sphere and enables students to investigate possible nets.

Practice 4.2B

1 and **2** allow students to practise isometric drawing. In question 2, students can control the level of challenge, depending on how they configure their four cubes.

3 asks students to draw plans and elevations for three solids. Students can first create the shapes out of cubes before going ahead with the drawings.

4 is the reverse of question 3 and is significantly more challenging. Cubes will definitely be helpful but, even then, some students will struggle and may need more support.

5 is a chance to review isometric drawings. The shapes are partially completed. Students are asked to use their spatial reasoning skills/spatial awareness to imagine the extra cubes added on to the diagrams. Some students may find it helpful to make the shapes.

6 interleaves work on sequences and square numbers whilst exploring plans and elevations further.

What do you think? 4.2B

1 is an open-ended task since there are many shapes that could be made.

2 is a chance to draw a shape with a triangular face on isometric paper, which is much more challenging than drawing a shape with rectangular or square faces.

3 explores a common misconception when moving from 2-D to 3-D shapes. Nets can be used for support.

Consolidate – do you need more?

This provides more opportunities to explore nets of 3-D shapes and accurate drawings, as well as consolidating work on plans and elevations. These questions can be used for homework or can be revisited later.

Stretch – can you deepen your learning?

1 is a discussion task that focuses on the fact that different shapes can look the same from certain directions.

2 is a tough question that challenges students' spatial awareness. They could use blank cubes for support.

3 gives some learners a chance to explore nets of a wider range of shapes.

4 The nets deliberately have square faces rather than rectangular faces. This makes the task much more challenging. Students could be challenged to come up with further valid nets.

Reflect

This question asks students to evaluate the advantages and disadvantages of different forms of representation.

4.3 Surface area (1)

Small steps
- Find areas of 2-D shapes
- Find the surface area of cubes and cuboids

Rationale

In this chapter, there is the opportunity to review work from Books 1 and 2 on the area of 2-D shapes, since much of this will underpin the work on surface area in the rest of this chapter and the next.

When looking at the area of 2-D shapes, students will review all the shapes they have met so far, including circles. There is ample opportunity to consolidate this, as well as to explore compound shapes and more challenging examples.

The chapter focuses on the surface area of cubes and cuboids. Students should be encouraged to appreciate the congruence of the opposite faces so that they can reduce the working needed to find the total surface area. Students may find it helpful to use nets.

Things to look out for	Interleaving opportunities
• Recalling formulae and using them correctly may be a problem for some students (for example, mixing up the area and circumference of a circle). • When working on reverse problems, where the area is given, students may need help to set up an equation and to think about the steps needed to solve it.	• The first part of this chapter provides a chance to review work on area from Books 1 and 2. • There are chances to interleave work on measures and converting units. • Other areas of maths, such as algebraic equations, ratio and using money, are also revisited within the questions.

Models and representations
- Everyday objects such as cereal boxes, tins and their labels, and other packaging should be used to give students a chance to explore nets in a hands-on manner.
- Geoboards can be used to explore 2-D shapes and area properties.

Behind the questions

Are you ready?

1 invites student to recall the definition of area. They can simply count squares at this stage but the formulae will soon be reviewed.

2 reminds students about the parts of the circle, which will be required later in the chapter.

3 is a chance to revise rounding, which will be needed when working with circles.

4 reviews solving equations since this will be required when solving problems that involve working backwards.

5 reminds students of basic unit conversion.

Practice 4.3A

1 is a chance to review finding areas of a variety of 2-D shapes, whilst question 2 adds circles into the mix.

3 offers more of a challenge by introducing a variety of compound shapes.

4 offers a chance to address the common error of using the slant height rather than the perpendicular height when more information than needed is given in a triangle.

5 requires students to work backwards. Some parts can be answered by inspection but students could also be asked to set up an equation in order to consolidate this skill.

6 would work well as a class discussion task to see how many different ways it can be done.

7 and **8** involve working backwards from the given area and provide another chance to review algebra skills.

9 reviews algebra skills by asking students to write an expression for the area of each shape.

10 challenges students to tackle a multi-step problem using the content from this section. There is a lot of information given, and the question is deliberately not scaffolded.

What do you think? 4.3A

1 Students compare the area of the two shapes. They could sketch the shapes and conjecture which would have the greater area before completing the calculations.

2 to **4** extend students with more challenging problems involving circles and part-circles. Question 4 is particularly challenging and needs careful thought. Asking students what they *can* work out may be helpful as they often struggle to get started in these types of questions.

5 invites students to think about the formula for the area of a trapezium and the relationships between the lengths in order to give an area of 30 cm^2. Students could think about ways of generalising their findings.

Practice 4.3B

1 to **3** involve finding the area of a cuboid. Students should be encouraged to think carefully about how to set out their working in order to avoid errors (such as missing a pair of sides and repeating another).

4 is a chance to develop reasoning skills by trying to explain why Ed would be wrong. Students could be prompted to think about the relationship between the original surface area and that of the stacked shape.

5 introduces an open box so that students don't get into a fixed routine when finding the surface area of a cuboid. This needs a bit of thought about the most efficient way to find the area (base and 4 identical sides).

6 introduces algebraic notation. Further questions are included in the "What do you think?" section.

7 and **8** involve solving multi-step problems. Question 7 is a bit more scaffolded whilst question 8 requires students to work through the question for themselves.

What do you think? 4.3B

1 uses more complex algebraic notation. Part b generalises the formula for the surface area of a cuboid.

2 is a working-backwards problem which is a bit more challenging than those met in the previous section. Students should be encouraged to set up an equation in terms of x as it will enable them to see a way forward. You could have a discussion about the usefulness of adopting algebraic approaches.

3 is an open-ended investigation. Students can explore this by using cubes and possibly by drawing their shapes on isometric paper, which will enable them to review the previous section.

Consolidate – do you need more?

This provides a variety of questions to consolidate finding areas of 2-D shapes and surface areas of 3-D shapes and includes some multi-step problems towards the end. This can be used for review or for homework and provides additional examples for those students who would benefit from them.

Stretch – can you deepen your learning?

1 is another chance to develop and use algebra skills. This question involves setting up an equation with brackets, which can be solved to find x.

2 interleaves work on ratio and requires several steps in order to find the answer.

3 is a further opportunity to make use of algebra skills from earlier in Book 3. These questions can be used to highlight the effectiveness of using algebraic methods.

4 Students need to consider the cubes that can be made from an A4 piece of paper. They need to consider different configurations of the net and find the one which would maximise the surface area.

Reflect

1 invites students to reflect on the area formulae for the different shapes they have met.

2 encourages students to explore the general formula for the surface area of a cuboid.

4.4 Surface area (2)

Small steps

- Find the surface area of triangular prisms
- Find the surface area of a cylinder

Rationale

The work of the previous chapter on surface area is extended here to include triangular prisms and cylinders.

Students will have already explored the nets of triangular prisms in depth. Therefore, they will be aware that each net is made up of two triangles and three rectangles. Some further exploration may be needed to explicitly link the lengths of the sides of the triangles with the widths of the rectangles. This is something students can find challenging so would be worth exploring using nets in the classroom.

For many, this may be the first exploration of nets of cylinders. Tins with their labels can be extremely useful for showing the relationship between the circumference of the circular base and the width of the curved surface. Students may need to set out their working in small steps initially rather than using the formula straight away.

Things to look out for	**Interleaving opportunities**
• Students can find it hard to visualise the faces of the prisms; 3-D objects and software can help. • There is scope for confusion when working with cylinders since the formula for the surface area contains within it the formulae for the area and for the circumference of a circle; so care must be taken here.	• Finding areas of individual faces reviews work from Books 1 and 2. • Students also build on the work from the previous chapter. • Other areas of maths, such as algebraic equations, ratio and using money, are also revisited within the questions.

Models and representations

- Everyday objects such as cereal boxes, tins and their labels, and other packaging should be used to give students a chance to explore nets in a hands-on manner.
- Folding geometry kits can be used to show students how nets and the 3-D solids fit together.

Behind the questions

Are you ready?

1 reviews finding the area of various shapes.

2 recalls how to find the circumference of a circle, since this will be required later in the chapter.

3 reminds students of leaving answers in terms of π when finding the area of a circle, whilst question 4 is an opportunity to review rounding.

5 makes connections between nets and the 3-D shapes they represent.

Practice 4.4A

1 asks students to draw the net of a prism, and to identify and label the lengths, before finding the surface area. This might be a method they choose to use for the remaining questions.

2 and **3** allow students to practise finding the surface area of a triangular prism. Question 3 has an extension which invites students to explore how changing one dimension might affect the surface area.

4 reminds students of the importance of paying attention to units and ensuring that all the lengths are in the same unit before carrying out any calculations.

5 offers an opportunity to practise finding the surface area and interleaves work on percentages from Book 2.

6 is a working-backwards question in which a missing length needs to be found. Students could do this intuitively or by setting up an equation. Different approaches could be explored and discussed.

What do you think? 4.4A

As well as practising finding the surface area, question 1 gives an opportunity to recall and use knowledge of ratio. In question 2, students will develop their spatial reasoning by imagining the cube cut in half and considering the newly exposed faces that will form part of the new surface area. Question 3 builds on the relationship between regular hexagons and equilateral triangles. Students may find it helpful to sketch the hexagonal face and mark on the given measurements as well as draw the resulting prism.

Practice 4.4B

1 asks students to draw a net of a cylinder, and to correctly label it, before finding the surface area. Students may also find it helpful to draw sketches for the remaining questions.

2 and **3** offer further opportunities to practise finding the surface area of a cylinder. Question 3 asks for the answer to be left in terms of π.

4 is an application problem for which students need to read the question carefully, since the lid and base are not included as part of the plastic surface area. This should discourage students from routinely applying a formula without first reading the question.

5 The cylinders have their dimensions swapped over, and the question involves finding which has the bigger surface area. Students could be encouraged to conjecture which it will be and to explain why they think this before carrying out the calculation. This could also be an opportunity to review estimating.

6 reviews percentage work from Book 2 in the context of the surface area of a cylinder.

7 and **8** provide more complex contexts for finding surface area. For question 7, using a real tin and demonstrating the overlap would be helpful.

What do you think? 4.4B

1 is a working-backwards question in which both dimensions are given in terms of x. This will enable students to apply their algebra skills to a more challenging question.

2 and **3** introduce more complex variations of cylinders. Ensure that students appreciate that the surface area of a semicylinder is not simply half that of a cylinder (as it would be for volume). Colour coding or labelling the different parts in question 3 can help to make sure that no part of the shape is missed.

Consolidate – do you need more?

This exercise provides consolidation questions on finding the surface area of prisms and cylinders. There are straightforward questions as well as some context-based ones. These can be used for review or homework.

Stretch – can you deepen your learning?

1 Although the chapter focuses on triangular prisms and cylinders, this question introduces a slightly different prism and allows students to revisit finding the area of a trapezium.

2 is challenging. Students should be encouraged to colour code or label the different faces in view so that they do not lose track of them. There are several parts to this question, and the need for clear working should be emphasised.

3 can be demonstrated easily by rolling up a piece of A4 paper. Students then need to consider the relationship between the circumference of the circle, its radius and the overall length of the paper.

Reflect

This question invites students to reflect on the complexities of the formula for the surface area of a cylinder and on why the circumference is needed. They are also asked to consider what may go wrong when finding the surface area of a cylinder, such as mixing up the formulae, using the diameter instead of the radius, etc.

4.5 Volume

Small steps

- Find the volume of cubes and cuboids
- Find the volume of other 3-D shapes – prisms and cylinders
- Explore volumes of cones, pyramids and spheres (H)

Rationale

Having spent much of this block exploring the properties and surface areas of 3-D shapes, this last chapter moves on to looking at the concept of volume.

To begin with, students explore the volume of cubes and cuboids. These can be made from linking cubes, in order to make the connection between the dimensions of the cuboid and the method for finding the volume. The idea of splitting a shape into layers or slices is utilised when explaining how to find the volume of a cuboid. This idea is then extended to looking at other prisms and cylinders.

The final section offers students an opportunity to explore the volumes of cones, pyramids and spheres. This is a Higher step and should only be tackled if students are confident with the earlier work.

Things to look out for	Interleaving opportunities
• It can be very easy for students to mix up the formulae for the surface area and the volume of a cylinder as they contain many of the same variables. • In the final section, the focus should be on volumes rather than surface areas as too many different formulae may lead to confusion.	• Finding areas of individual faces reviews work from Books 1 and 2. • There is also the opportunity to review the algebra skills needed to set up and solve equations. • Number topics from Books 1 and 2, such as ratio and percentages, are revisited.

Models and representations

- Cubes can be used to demonstrate the concept of volume and to show how to find the volume of a cuboid.
- Actual 3-D shapes, such as those made from wood, can be used in class to help students to visualise them.

Behind the questions

Are you ready?

1 asks students to find the area of a variety of shapes whilst question 2 provides a reminder of rounding.

3 and **4** review algebra skills, including substitution and solving simple equations. Both will be required when working with volume formulae.

Practice 4.5A

1 allows students to imagine the cubes that make up a shape when finding its volume. They can simply count the cubes but should be encouraged to find a more efficient method (which leads to the formula for volume).

2 provides practice at finding the volume of cuboids.

3 brings in a missing length and invites students to think about different approaches to finding the value of x.

4 reminds students of nets of cuboids and how to recognise their length, width and height from the diagram.

5 is an opportunity to review litres and unit conversions, which will be used again later in the chapter.

6 provides a selection of questions in which students need to work backwards to find missing dimensions. They should be encouraged to think about different approaches.

7 reminds students of the commutativity of multiplication, for example that $l \times w \times h = w \times h \times l$.

8 is an opportunity to apply knowledge of volume to a simple question involving a shipping container.

9 explores the effect of a percentage change in length on volume. Although repeated change has not been covered yet, students should see that a 10% increase in length does not lead to a 10% increase in volume.

What do you think? 4.5A

1 is a multi-step problem in which no scaffolding is given. Students need to think about what they know, what they can find out and how this will help them to answer the question.

2 introduces the concept of rate of flow as well as the use of litres as a unit of volume. Students will need to convert from minutes to hours and minutes. Therefore this question offers plenty of interleaving opportunities.

3 is a good discussion task that students could work on in pairs. Again, they need to consider what they are able to work out from the information in the question and how this will help them.

Practice 4.5B

1 to **3** offer plenty of practice at finding the volumes of prisms and cylinders and leaving answers in terms of π.

4 to **6** invite students to solve problems in a variety of contexts.

7 is a selection of working-backwards questions.

8 reintroduces the semicylinder. Students could be asked to compare the relationships between the surface areas and volumes when going from a cylinder to a semicylinder.

What do you think? 4.5B

1 is an excellent opportunity to practise problem solving: it involves fractions, litres as a unit of volume measurement and working backwards to find the height when given the volume.

2 and **3** offer opportunities to practise more challenging, multi-step problems. In question 2, students may need help to identify the radii of the cylinders.

Practice 4.5C

1 and **2** give students the chance to practise finding the volume of a variety of solids, including a compound one.

3 allows students to address misconceptions and could be an opportunity to review estimating and reasoning.

4 is a simpler problem-solving task that reviews percentages from Book 2.

5 provides working backwards questions with more complex formulae, which may be a chance to review algebra skills such as solving equations by balancing.

What do you think? 4.5C

1 and **2** are both challenging questions, and question 2 needs good algebra skills. Both can be used to develop resilience and strategies for problem solving, for example "write down what you know", "what can you work out?", "what information do you need?". Such questions can really encourage students to get started!

Consolidate – do you need more?

This exercise provides consolidation questions on finding volumes and includes problems set in context.

Stretch – can you deepen your learning?

1 requires unit conversions and working backwards to find the diameter (not radius!) of a cylinder.

2 reviews work on volume, ratio and surface area, allowing an opportunity to consolidate these skills.

3 is a working-backwards, multi-step, problem-solving question.

4 reviews ratio and fractions in the context of volume.

5 addresses a possible misconception about a half-full cone.

Reflect

Question 1 invites students to explore items in the home and to think about when the different units of volume are used. For those who have worked on the Higher section, question 2 asks students to reflect on the distinction between capacity and volume, which confuses many people. This is an opportunity for discussion.

5.1 Constructions so far

Small steps

- Draw and measure angles (R)
- Construct and interpret scale drawings (R)
- Construct triangles from given information (R)

Rationale

This chapter reviews students' learning about constructions from Books 1 and 2 in order to prepare them for the more complex material in the following chapters. Encourage students to use estimation to consider whether their diagrams look reasonable.

Things to look out for	Interleaving opportunities
• Incorrect positioning of rulers and protractors. Model their use carefully to remind students of the need for accuracy. In particular, encourage estimation before measuring/drawing an angle to ensure that students use the correct scale on their protractors. • Difficulty using pairs of compasses. Students may need to be reminded to hold the compasses at the top rather than at the arm. • Blunt pencils make diagrams far less accurate, so you may need a large supply of sharpeners and erasers! • Students may need a reminder about how three-letter notation is used to name sides, angles and shapes.	• Ratio notation is revisited when looking at scale diagrams. • Students will use the basic skills in this chapter as the basis for the rest of this block. • Construction of triangles can be practised in Block 4, when looking at nets, and in Block 12, when looking at enlargement and similar triangles. • Conversion of metric units. • Much of the work on construction does not interleave easily into other areas of the curriculum, so it might be worth setting aside specific time to practise it, e.g. in starter activities.

Models and representations

- As illustrated in the Student Book, many real-life objects can be manipulated to produce angles. These are useful for developing estimation skills and students' "feel" for the size of an angle, as well as for reminding them that angles are measures of turn rather than a mathematical construct based on calculation.
- Actual scale drawings of real places such as the school or maps of the local area(s) may spark students' interest and help them to visualise scaling by factors such as 100 or 1000.

Behind the questions

Are you ready?

Questions 1 and 2 focus on reminding students how three-letter notation works for naming angles; the arithmetic involved in the questions is secondary. Question 3 reminds students of the word "obtuse", and you could extend this to bring in all of the other vocabulary associated with angles. Question 4 pre-empts possible issues with the misuse of a protractor; a discussion around this might help to reduce errors later.

Practice 5.1A

1 practises drawing angles in a variety of orientations. Students could rotate their exercise books to make this easier, if desired.

2 Students measure both acute and obtuse angles; they are specifically instructed to estimate in part a. You may wish to cover comparison with a right angle as a tool for estimation (e.g. using the corner of a piece of paper) and encourage students to rotate the diagrams to help them to see this.

3 compares the possible methods for drawing reflex angles. It is easier to check by measuring the complement to 360°.

4 uses the fact that opposite angles in a parallelogram are equal as a vehicle for more measuring practice, and it also reminds students of the property of co-interior angles.

5 practises using a scale drawing. A brief discussion may be needed as a reminder before students tackle this.

6 requires students to work backwards from a given diagram.

7 builds on questions 5 and 6 by looking at map scales.

8 asks students to compare different scales for the same object.

9 is designed for discussion. You may wish to compare them with the height of an average adult person.

What do you think? 5.1A

1 Students record their reasoning about the use of a protractor as practised in the exercise.

2 and **3** provided more opportunities to discuss scales.

Practice 5.1B

1 provides practice at the AAS construction. Here, and throughout this exercise, students can check each other's accuracy.

2 models and provides practice at the SSS construction.

3 models and provides practice at the SAS construction.

4 discusses the meaning of the abbreviated names.

5 and **6** provide further practice without the scaffolding of the required technique being named. **6** requires understanding of three-letter notation.

7 The quadrilaterals can be drawn by considering pairs of joined triangles. This could be extended to consider other polygons.

What do you think? 5.1B

1 considers the need for side lengths and links forwards to similarity (Block 12).

2 checks that students appreciate that the lengths of the sides need to be sufficient for the sides to meet.

3 extends to the construction of a regular hexagon.

Consolidate – do you need more?

Here the key skills are revised. As this topic is more difficult to interleave than most, it may be worth using this as a revision exercise later in the year.

Stretch – can you deepen your learning?

1 to **3** explore bearings as another context for practice. Bearings are studied more formally in KS4. Other than the definition, no extra learning is involved.

Reflect

These questions cover the key points of the chapter. Encourage students to show their understanding in their own words.

5.2 Introducing loci

Small steps

- Locus of distance from a point
- Locus of distance from a straight line/shape

Rationale

In this chapter, students consider the meaning of the word "locus" as a set of points that satisfy a given rule. The suggested approach is to use coins and counters to help to support their visualisation of the paths and then to construct these accurately using geometrical equipment. It is worth investing sufficient time in the early stages for students to become familiar with the term, as it is likely to be new. Good motor skills are needed to construct loci accurately. Time spent constructing patterns using, for example, overlapping circles is a good way of helping students to develop these skills.

Things to look out for

- Difficulty with the unfamiliar language. To help students to remember it, persist with the use of precise mathematical language throughout the chapter and try to introduce the concept of a locus when studying other blocks.
- Difficulty using pairs of compasses. Students may need to be reminded to hold the compasses at the top rather than at the arm.
- Blunt pencils make diagrams far less accurate, so you may need a large supply of sharpeners and erasers!

Interleaving opportunities

- The skills of Chapter 5.1 are built upon, both in the use of pairs of compasses and for drawing diagrams to a simple scale.
- Construction does not interleave easily into other areas of the curriculum, so it might be worth setting aside specific times to practise this, for example in starter activities. Key terms such as "locus" can be reinforced in the same way.

Models and representations

- Coins and counters are useful for both generating the shapes of the loci studied in this chapter and to illustrate the need for the use of geometrical equipment to produce accurate, rather than approximate, sets of points. Suggested activities are provided in the exercises, and you can vary the distances and loci, as required.
- "Human geometry" is a fun alternative that can be used in a large room or the playground. Students can be provided with metre sticks or tape measures and asked to stand a certain distance, e.g. 2 m, from a point or a wall (see question 2 of Practice 5.2B).

Behind the questions

Are you ready?

Question 1 provides proficiency practice at using a pair of compasses and question 2 checks that students remember the difference between the radius and diameter of a circle. Question 3 provides more practice at using a pair of compasses. You could invite students to create their own designs, or to see how small they can accurately produce the one given, to promote the development of their motor skills and accuracy.

Practice 5.2A

1 Students can use different distances to reach the general result that the locus of points at a certain distance from a fixed point is a circle. You could also discuss using the corner of a desk to produce a quadrant.

2 gives practice at interpreting the formally phrased request for a locus that is a circle, wit part c introducing the set of points inside the circle. Discuss the need for precise language when describing loci.

3 develops the ideas from question 2 to look at loci defined by more than one set of requirements. You could extend this by shading some regions and asking students to describe them.

4 and **5** develop the ideas introduced in question 3.

6 links to the construction of triangles in the previous chapter. It also links forwards to Pythagoras' theorem in Block 11.

7 looks at restricting the domain of the loci within a shape. The last part allows students to be creative by making up their own loci.

What do you think? 5.2A

Question 1 provides a context for using loci to define regions and explores why a region inside a shape might be more relevant or practical than one outside. The advantages and disadvantages of the circular regions are explored in question 2, which provides plenty of opportunities for skills practice in this investigative context.

Practice 5.2B

1 mirrors question 1 of Practice 5.2A, but now looks at the locus of points at a certain distance from a line rather than a fixed point. The aim at this stage is to engender understanding rather than to record results.

2 This activity, with the numbers suitably adjusted, could be carried out practically before answering the question, if space allows. Loci from both a point and a line are covered, including the regions they produce as well as the paths.

3 tackles a common misconception associated with drawing the locus of points that are a fixed distance from a line. This is worth class discussion. It is easy to prove that the first alternative is wrong by measuring. Students may be interested to learn that the locus can be called a "stadium" (link this to the shape of a running track) or a "disco-rectangle".

4 develops this to the locus outside a shape. Students should consider the line segments and vertices separately.

5 repeats question 4 in the context of a triangle. You could include other triangles to encourage further practice.

6 looks at regions again, this time inside a shape.

What do you think? 5.2B

These two questions provide opportunities for exploration of the activities covered in the chapter. This can be refined by including other obstacles, such as posts and walls, that would restrict the dog's movement.

Consolidate – do you need more?

This skills-practice exercise could be saved and used for revision later in the year, if appropriate.

Stretch – can you deepen your learning?

1 and **2** form an open-ended investigation that extends the ideas of the chapter. A similar idea is to consider the location of security cameras inside/outside buildings made from various rectilinear shapes.

3 Encourage students to use a pair of compasses to consider the path of P as the box rotates. This could be extended by looking at rectangles in different initial orientations and by using other shapes.

Reflect

This reflection compares and contrasts the loci met in this chapter.

5.3 Perpendiculars

Small steps

- Locus equidistant from two points
- Construct a perpendicular bisector
- Construct a perpendicular from a point
- Construct a perpendicular to a point

Rationale

As in the previous chapter, the locus is introduced by first thinking about individual points, generalising and then moving on to the formal construction. Students need to be familiar with both ways of phrasing the instructions, i.e. "Construct the perpendicular bisector …" and "Find the set of points equidistant from …", and they should be able to interpret the latter in context. Students who completed all the Higher steps in Book 2 will have met the perpendicular bisector idea before, but this will be new learning for many. This is developed to include the other standard perpendicular constructions, which are linked to the perpendicular bisector.

Things to look out for

- Difficulty with language – breaking down the word "equidistant" into two parts helps to make its meaning clear, but the word needs explanation when first seen.
- Issues with accuracy, as discussed in Chapters 5.1 and 5.2. Again, plenty of practice is needed to develop accurate construction skills. You can use this as an opportunity to check measuring skills.
- Students may need guidance to appreciate that the locus of points equidistant from two fixed points and the perpendicular bisector of the line segment joining the two points are the same thing. Ensure that this is discussed.
- Students sometimes find the construction of the perpendicular bisector difficult if the line segment/points are not horizontal. The exercises explore various orientations, so students may find it helpful to rotate their books.

Interleaving opportunities

- The skills from Chapters 5.1 and 5.2 are developed.
- Constructing perpendicular bisectors in some shapes will lead to observations about line symmetry. This will be linked to the diagonals of a rhombus in Block 9.
- This topic does not interleave easily into other areas of the curriculum, so it might be worth setting aside specific times to practise it, for example in starter activities.

Models and representations

- Coins and counters are again used to generate the shape of the locus before moving on to formal constructions using a pair of compasses.
- Paper folding is useful for demonstrating the position of a perpendicular bisector. In particular, this might help students to carry out the construction when the two points or the line segment are not presented in a horizontal line.

Behind the questions

Are you ready?

Question 1 reminds students of the meaning of the word "perpendicular". Questions 2 and 3 check students' understanding of the word "locus" and the difference between the locus of points equidistant from a point and from a line, all of which were covered in the previous chapter.

Practice 5.3A

1 explores the idea of equidistance with coins and counters. Encourage students to look at various orientations, not just horizontally and/or vertically.

2 looks at finding a pair of points that are equidistant from two points. This is linked to Example 1 but demonstrates that full circles are not needed.

3 Students are asked to construct a set of points that are equidistant from the endpoints of a line segment, so that they appreciate that the resulting locus is both perpendicular to the line segment and the same distance from both endpoints. Working through the steps in the question should help students to understand why the standard construction works.

What do you think? 5.3A

Students should conclude that one pair of arcs that cross "above" and "below" the line (depending on orientation) should be sufficient, which thus leads to the standard construction.

Practice 5.3B

1 walks students through the standard construction.

2 explores different placements of the line segment.

3 is fluency practice. You may wish to provide students with a selection of line segments or allow them to choose their own. You can discuss the relative difficulty of shorter/longer segments and their orientations.

4 When they are fully competent, students can move on to the construction of a perpendicular line through a point on a line segment. Notice that this is not presented as a different "new" construction to learn, but as a special case of using the perpendicular bisector.

5 provides practice at drawing the construction introduced in question 4.

6 scaffolds the idea of extending a line segment in order to construct the perpendicular to it through a point close to or at the end of the line.

What do you think? 5.3B

Question 1 challenges students to adapt the construction of the perpendicular bisector to construct the perpendicular from a point to a line. You may wish to discuss this as a class instead, as the answer is provided in question 3 alongside some suggested practice. These questions are split up by question 2, which looks at the location of the centre of a circumcircle; this is developed in question 3 of the "Consolidate" exercise.

Consolidate – do you need more?

This provides more practice at all the constructions in the chapter. Question 3 uses the circumcircle as a vehicle for this, and builds on question 2 in the second "What do you think?" section. This activity could be used as the basis for a nice classroom display. Students should be encouraged to avoid drawing small triangles as this makes the construction more difficult.

Stretch – can you deepen your learning?

1 is accessible to all and explores the constructions using dynamic geometry software.

2 challenges students to understand a construction presented in formal language.

3 links to straight-line graphs, in particular to midpoints and gradients and the fact that the product of the gradients of perpendicular lines is –1.

Reflect

This encourages students to see the relationships between the constructions covered in this chapter rather than seeing them as three different techniques to be learned.

5.4 More loci

Small steps

- Locus of distance from two lines
- Construct an angle bisector

Rationale

This short chapter looks at the construction of an angle bisector and links this to the locus of points equidistant from two lines in the same way as the perpendicular bisector was introduced in the previous chapter. Again, students who completed all the Higher steps in Book 2 will have met the angle bisector before, but this will be new learning for many. There are alternative ways to construct an angle bisector other than the one modelled here, one of which will be covered in Block 9, but to avoid possible confusion only the more common approach is demonstrated in this chapter.

Things to look out for

- Issues with accuracy as discussed in the earlier chapters in this block. As always, plenty of practice is needed to develop accurate construction skills. You can use this opportunity to check that students are proficient at drawing and measuring angles.
- Students may need guidance to appreciate that the locus of points equidistant from two lines relates to bisecting the angle rather than the lines, but the practical experience should support this.
- Mixing up the constructions of the perpendicular and angle bisectors. Relating these to what they represent and what distances they are trying to keep equal may help with this.
- Students will need to appreciate that the shortest distance between two parallel lines is the length of the perpendicular line segment between them.

Interleaving opportunities

- The skills introduced in earlier chapters in this block are developed. In particular, the loci in Chapters 5.2 and 5.3 are explicitly revisited.
- The angle bisector will be linked to the construction of a kite in Block 9.
- As with the rest of this block, this topic does not interleave easily into other areas of the curriculum, so it might be worth setting aside specific times to practise it, for example in starter activities.

Models and representations

- Coins and counters are again used to generate the shape of the locus before progressing to formal constructions using a pair of compasses.
- Paper folding is useful for demonstrating the position of the angle bisector.
- "Human geometry" can be used. Students holding measuring equipment, e.g. tape measures, could stand at points equidistant from two lines.

Behind the questions

Are you ready?

Question 1 practises the use of pairs of compasses and revisits the learning of the previous chapter, and question 2 checks protractor skills as revised in Chapter 5.1. Question 3 revisits the idea of a locus from a point as covered in Chapter 5.2. This chapter will use these ideas in some of the more complex problems that look at regions that satisfy a set of criteria.

Practice 5.4A

1 explores the idea that the locus of points equidistant from two lines bisects an angle. You could extend this to looking at other lines, for example using pencils:

2 works through the formal construction, which students then practise in questions 3 and 4 using a variety of sizes and orientations of acute angles.

5 Students continue to practise the construction using obtuse angles. You can add in more skills practice here if needed, or go on to question 6, which uses the incircle as a vehicle for this.

7 You could show that extending the bisector of the complementary angle to 360° will bisect the reflex angle:

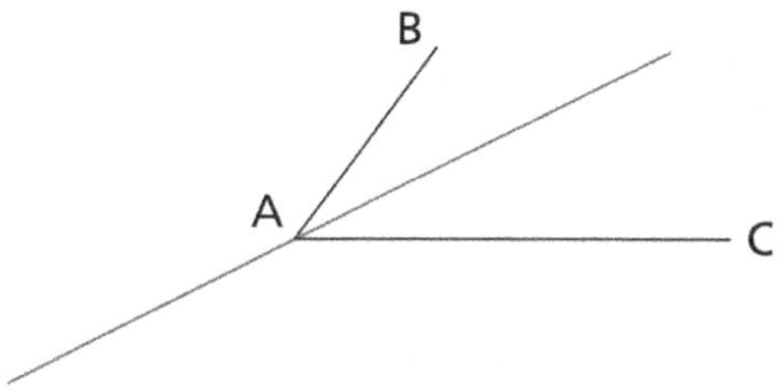

8 highlights the distinction between angle bisectors and perpendicular bisectors.

What do you think? 5.4A

1 is designed to promote discussion, for example, "How can you measure the perpendicular distance of the points to the left of A from AB and AC?"

2 looks at a common error.

3 links to the loci of Chapter 5.2. Students could create their own questions for further practice.

Consolidate – do you need more?

This exercise provides further skills practice. Question 3 includes both angle and perpendicular bisectors.

Stretch – can you deepen your learning?

1 combines drawing the angle bisector with familiar skills to construct angles of various sizes without a protractor.

2 The angle bisector theorem makes a nice link with ratio. Formal proof of this is not required.

Reflect

This checks understanding of the vocabulary of the chapter and helps students to compare and contrast the different bisectors that they have learnt to construct.

5.5 Congruence

Small steps

- Identify congruent figures
- Explore congruent triangles
- Identify congruent triangles

Rationale

Students explore the nature of congruent figures in order to appreciate that different rotations, reflections and translations essentially leave a shape unchanged. This is in contrast to enlargement and similarity, which are covered in Block 12. Students then explore the conditions for congruent triangles via construction. They should be able to use the conditions to show that triangles are congruent but, at this stage, it would be helpful to have these provided rather than expect students to memorise them. More formal proof of congruence can be left until KS4.

Things to look out for	Interleaving opportunities
• Not recognising that shapes that have been reflected are congruent. Include these in your examples. Students could also copy, cut out and check congruence physically. • Difficulty with remembering all the conditions for congruent triangles. Provide these at this stage, and encourage students to look at the information given and their list and decide which one(s) would be most sensible to check first.	• Construction of triangles from Chapter 5.1 is revisited to establish the conditions for congruent triangles. • The difference between congruence and similarity will feature in Block 12. The conditions for congruence will be needed for some of the questions here and also in Block 9, where geometric deduction is explored further.

Models and representations

- Geoboards are very useful for comparing shapes, e.g. "How many different triangles can you make in a 2 by 2 section of the board?" Dotty paper can be used as an alternative (see question 7 of Practice 5.5A).
- Students can draw and cut out shapes, flip them over and turn them so that they fit on top of each other to show congruence.
- Pattern blocks are sets of congruent shapes that can be joined edge-to-edge in twos or threes to produce other shapes. Depending on where these are joined, the resulting shapes may or may not be congruent. This provides a good starting point for a discussion about the concept of congruence.

Behind the questions

Are you ready?

Question 1 reviews the construction of triangles in preparation for the upcoming investigation into the criteria for unique/non-unique triangles. Questions 2 and 3 review simple triangle properties and names alongside the meanings of geometric notation that will be needed for identifying pairs of congruent triangles.

Practice 5.5A

1 Students explore the idea of congruence on squared paper. Compare responses including, for example, rotations of 45°.

2 and **3** increase the level of difficulty by moving away from rectilinear shapes. Students could cut out copies of the shapes to check their answers physically, if necessary.

4 checks understanding of geometric notation. You could give each point a label and ask students to identify the corresponding points in the two shapes.

5 provides further practice with more complex shapes.

6 requires students to understand properties from descriptions in words.

7 is an investigation which explores congruent triangles. You could also look at other sizes of grid. Dotty paper or geoboards are useful for support.

8 asks students to identify corresponding sides and angles.

What do you think? 5.5A

1 tests geometric notation, but also introduces the conditions for congruent triangles.

2 tests the construction of triangles and allows students to start to formally explore the conditions for congruent triangles, which are then built on in the following text, examples and exercise. This can be used as a basis for a very useful class discussion as to why the conditions hold.

Practice 5.5B

1 looks at identifying corresponding sides in congruent triangles using the SSS condition. You could also write the answers in the form

Triangles $\begin{matrix}\text{ABC}\\\text{RPQ}\end{matrix}$ are congruent.

2 and **3** are similar to question 1 but use the SAS and AAS conditions. The focus of these three questions is on understanding rather than checking congruence.

4 extends the concept of congruence by asking students to identify the correct condition.

5 provides scaffolding to guide students towards writing a formal proof of congruence by asking about the vertically opposite angles first.

6 introduces the idea of a common side.

7 looks at the need to find a missing angle using the angle sum property of a triangle.

8 helps students to see why "two sides and an angle" is not a sufficient condition for congruence if the angle is not the included angle between the two sides.

What do you think? 5.5B

1 explores the RHS condition.

2 makes students aware that RHS is not the only possibility for congruence of right-angled triangles.

Consolidate – do you need more?

This provides further practice at identifying congruent shapes and the conditions necessary for triangles to be congruent.

Stretch – can you deepen your learning?

1 explores congruence through triominoes and tetrominoes. Squared paper, or squares from a set of pattern blocks, can be used for support.

2 to **4** look at more formal proofs that a pair of triangles are congruent using properties of shapes and angles in the style of GCSE questions. There is often more than one approach, and these could form the basis of a good discussion. You could also look at actual past paper questions, if appropriate.

Reflect

These questions could form the basis of a class display.

6.1 Working with numbers

Small steps

- Integers, real and rational numbers
- HCF and LCM (R)
- Understand and use surds (H)

Rationale

In this chapter, students look at different types of numbers and their position in the number system. The first part of the chapter focuses on integers, real numbers and rational numbers. Students should already be confident at identifying and explaining the key features of integers but may not be familiar with the terms real and rational numbers. Although students do need to know that real numbers can be either rational or irrational, they do not need to know about imaginary numbers. However, you could inform them that this is something they may find out about if they choose to study maths beyond GCSE. In the second part of the chapter, students revisit the concepts of highest common factor and lowest common multiple. Some students may not have used factor trees to work out the prime factors of numbers previously; this is a good opportunity to introduce this method and discuss its pros and cons. In the final part, students are introduced to the idea of a surd. They should be able to recognise a surd and carry out basic simplification of surds. There is no need to extend beyond this at KS3.

Things to look out for	Interleaving opportunities
• Students often confuse factors and multiples; linking the word "multiples" to "multiply" can help to overcome this barrier. • Some students may confuse HCF and LCM and think of them as HCM and LCF, respectively. They should be prompted to realise why it is not possible to find the highest common multiple of any pair of numbers, and why the lowest common factor would be the same for any set of numbers. • When working with surds, students may write the wrong number under the square root symbol, for example by writing $\sqrt{50}$ as $2\sqrt{5}$ instead of $5\sqrt{2}$. To avoid this mistake, students should be encouraged to write down each step of their working.	• The equivalence of fractions, decimals and percentages will be used throughout when working with rational and irrational numbers. • Place value can be used to support understanding of rational numbers, as shown in Example 1. • Students are reminded of prime factor trees from Book 1; they use prime factorisation to calculate the HCF and LCM of a set of numbers using a Venn diagram. • Square numbers are revisited when simplifying surds.

Models and representations

- Venn diagrams can be used to show where each type of number lies within the number system, and provide a way of showing prime factors that can be used to find the HCF and LCM of two or more numbers.

Behind the questions

Are you ready?

The focus here is on basic factors and multiples as the rest of the learning in this topic may be new for students. Notice how the numbers used in questions 1 and 2 are identical; this is to draw attention to the difference between factors and multiples to try to prevent any confusion later in the chapter.

Practice 6.1A

1 to 4 provide opportunities for students to practise finding and identifying integers, real numbers and rational numbers. Students should be encouraged to explain their reasoning in each case.

5 and **6** require students to use their knowledge of fraction and decimal equivalents to explain why each of the numbers given is rational. In question 5b, students could also consider whether *any* recurring decimal is irrational.

What do you think? 6.1A

These two questions require students to think more deeply about integers, real numbers and rational numbers. They should realise that although the numbers presented in questions 1a, 1c and 2 don't immediately look like integers, in fact they are, and students should be able to explain why this is the case. They could be asked to generalise.

Practice 6.1B

1 Students might need reminding that factors come in pairs. They can use this fact to ensure that they don't miss out any factors. Systematic approaches to finding factors should be encouraged.

2 Prompt students to consider whether or not 10 multiples were needed and, if not, when they would stop when listing the multiples of each number.

3 Here students use a Venn diagram to calculate the HCF and LCM. Students should consider when the methods used in questions 1 and 2 are more appropriate and when a Venn diagram is more efficient.

4 and **5** Here students can choose their preferred method for finding the HCF or LCM and should be able to justify their choices. They could be encouraged to compare their method and rationale with a partner.

6 and **7** extend students' knowledge of using a Venn diagram to calculate the HCF and LCM and allow them to demonstrate their understanding of these concepts.

9 requires students to calculate the LCM in context. They will first need to identify that it is the LCM of the numbers that is required and explain why this is the case.

What do you think? 6.1B

Question 2 focuses on the idea of two numbers having no prime factors in common (i.e. being coprime) and what this means about their HCF. Students could be stretched to consider how it affects their LCM.

Practice 6.1C

2 and **3** use facts that students already know to introduce some basic rules of surds.

4 and **5** enable students to consolidate their understanding of using the rules of surds to simplify expressions.

6 and **7** provide practice at simplifying surds. Notice how the surds in both questions are the same. In question 6, the focus is on identifying square factors and, in question 7, they then use these to simplify the surds. Students should realise that choosing the greatest square factor in question 6 would make the simplification in question 7 more efficient. This could be linked to the idea of using the HCF to simplify fractions.

What do you think? 6.1C

These three questions require students to think more deeply about surds by using them in different contexts and exploring a common misconception.

Consolidate – do you need more?

These three questions allow students to consolidate their understanding of finding the HCF and LCM.

Stretch – can you deepen your learning?

1 Students should be able to explain their reasoning for each statement using examples or counterexamples, where appropriate.

2 prompts students to identify the information they can find from a prime factor Venn diagram.

3 and **4** link the learning from this chapter to prior learning on conjectures and algebra.

Reflect

Students are encouraged to reflect on their learning by explaining in their own words what is meant by the terms "integer", "real number" and "rational number". They are also asked to define HCF and LCM and explain the difference between the two.

6.2 Estimation

Small steps

- Solve problems with integers
- Solve problems with decimals

Rationale

In this chapter, students look at estimation. The first part of the chapter focuses on estimating with integers and decimals and builds on students' knowledge from the number sense units in Books 1 and 2. Students should know that they need to round to 1 significant figure to estimate the answer to a calculation and should then be able to use this skill in context. They should be able to explain the benefits of estimating and how it can be used to sense-check answers. The second part of the chapter looks at error intervals; these were covered as a Higher small step in Book 2, so this topic will be new for some students.

Things to look out for	Interleaving opportunities
• Students are often tempted to work out the actual answer rather than an estimate. • Some students might round decimals to the nearest integer rather than to 1 significant figure and should be prompted to recognise their mistake. • Students may confuse the inequality symbols when writing error intervals. They should use their knowledge of rounding to identify their mistakes and appreciate which inequality symbols need to be used.	• Students will need to follow the correct order of operations when estimating the answers to calculations. • Students should be reminded of rounding to different degrees of accuracy in preparation for writing error intervals before focusing on rounding to 1 significant figure for estimation. • Students will encounter estimation in many different areas of the curriculum, e.g. area, money, algebra, etc.

Models and representations

- Number lines can be used as a tool to support understanding of both rounding and error intervals.

Behind the questions

Are you ready?

These three questions will ensure that students have the essential prerequisite knowledge for this chapter. In question 1, students recap rounding to 1 significant figure. Watch out for them rounding to the nearest integer in part e rather than to 1 significant figure. Question 2 provides practice at performing basic calculations involving numbers with 1 significant figure. Part d may cause some confusion; students could make links to money, e.g. "How many 50ps are there in £30?" to assist their understanding. Question 3 requires students to think about which numbers might round to 1000 to a given degree of accuracy.

Practice 6.2A

1 provides a scaffolded approach to estimation. Part a instructs students to round the numbers to 1 significant figure, and then they use their answers to estimate the value of each calculation in part b.

2 Notice that the calculations generally involve a single step so the focus can be placed on the rounding.

3 builds on question 2 by looking at multi-step calculations. Students may need reminding of the order of operations here. They may also be tempted to round midway through, e.g. in part b they should rewrite the calculation as $\frac{700+8000}{20}$, which simplifies to $\frac{8700}{20}$. Students may want to round 8700 to 9000 before continuing the calculation, and this could form the starting point for an interesting discussion about why we don't do that.

5 revises calculating the area of a variety of different shapes. In parts c to e, students should be encouraged to explain which lengths they need and which they don't, to demonstrate their understanding.

6 looks at currency conversion. This is a good opportunity to discuss the real-life benefits of estimation as it is often useful to roughly convert prices into different currencies.

7 provides an opportunity to revisit substitution. Students might think that parts e and f are equivalent, and they should use estimation to demonstrate that this is not the case.

8 Students could be stretched to consider which other averages they could estimate.

What do you think? 6.2A

These four questions provide opportunities for students to think more deeply about estimation. They are encouraged to consider whether an answer is an overestimate or underestimate depending on how the numbers in the calculation have been rounded. Question 3 addresses the misconception that may have arisen in question 3 of Practice 6.2A. Question 4 looks at an example where 1 significant figure isn't the most appropriate rounding. Students should be able to explain why it is that, when estimating roots, rounding to 1 significant figure isn't always the most sensible degree of accuracy.

Practice 6.2B

1 provides a scaffolded introduction to error intervals. Students first consider which values from a given list round to 4 before thinking about how this could be recorded more generally. They are asked to decide which is the correct way of using an error interval to represent numbers between 3.5 and 4.5 that all round to 4.

2 allows students to practise interpreting error intervals. For part b, they may wish to write several possible values for a before realising that they all round to the same integer.

3 Here students are required to form their own error intervals. Part a is scaffolded whilst, in part b, students are only given the degree of accuracy that the value has been rounded to. Students could be encouraged to draw a number line for support.

4 looks at how the error interval changes when the degree of accuracy changes. This is a good starting point for a discussion about why it's important that we know the degree of accuracy that a value has been rounded to in order to write the correct interval.

5 to **8** enable students to practise writing and interpreting error intervals in different contexts.

What do you think? 6.2B

Question 1 focuses on how errors in individual measurements can have a knock-on effect on the total length of an object. Question 2 looks at non-standard degrees of accuracy for rounding and prompts students to generalise and form an expression for the error interval given any degree of accuracy.

Consolidate – do you need more?

These five questions allow students to consolidate their understanding of rounding and estimating.

Stretch – can you deepen your learning?

1 Students should be able to explain why the answer in terms of π is more accurate.

2 introduces students to a different way in which rounded values can be given and then asks them to think about how these rounded values affect the error intervals for the perimeter and area of a square.

3 Here students could be encouraged to work out all possible values of the speed and then identify the endpoints of the error intervals.

4 Students should be prompted to recognise that 17.45 is 2.55 less than 20 and 42.55 is 2.55 more than 40, which means that 20 × 40 gives an overestimate because

$(20 - 2.55) \times (40 + 2.55) = (20 \times 40) - (20 \times 2.55) - 2.55^2$

5 Students might be tempted to think that this conjecture is always true and should be directed to look back to question 3 in the "Are you ready?" exercise to understand why it isn't.

Reflect

Students are encouraged to reflect on their learning by explaining in their own words how to estimate the answer to a calculation and the importance of error intervals in real life.

6.3 Solving problems with numbers

Small steps

- Work with directed numbers (R)
- Solve problems with integers
- Solve problems with decimals

Rationale

This chapter provides opportunities for students to consolidate their arithmetic skills whilst applying them to solving problems in a variety of contexts. Students need to be able to calculate with integers, decimals and directed numbers. In each question, students should carefully consider what they know and what they want to find out in order to help them to decide the steps needed. The use of pictorial and concrete representations can support this decision-making process.

Things to look out for	Interleaving opportunities
• Numerical errors are very common when performing basic calculations, and students should be encouraged to check their answers carefully to identify any mistakes. • Some students might incorrectly think that two negatives always make a positive and should be shown examples of where this isn't true to unpick this misconception. • Students might not be confident at finding complements to 1 when working with decimals. For example, they may think that the total of 3.5 and 4.7 is 7.12 because they don't recognise that 12 tenths is equivalent to 1.2.	• The four operations are used throughout this chapter, and students could be reminded of their work on number sense to enable them to choose efficient methods of calculation. • This chapter could be used to consolidate the previous one on estimation. Students could use estimation as a method of sense-checking their answers. • Students can solve problems involving many different areas of the curriculum, e.g. working with units of measure.

Models and representations

- Double-sided counters can be used to represent directed numbers.
- Bar models are useful for helping students to decide which calculation they need to perform.

Behind the questions

Are you ready?

These three questions ensure that students have the essential prerequisite knowledge for this chapter. Question 1 provides practice at performing calculations using a mental or written method whilst questions 2 and 3 are designed to remind students of the basic rules of working with directed numbers. As this chapter relies on students being able to perform these sorts of calculations efficiently, it may be worth spending some time modelling some appropriate methods. Estimation can be used to check answers, which will revise the work of the previous chapter.

Practice 6.3A

1 A common mistake here is to just give the amount of money Ali has rather than the total amount.

2 Dividing 365 by 45 gives an answer that rounds to 8. However, students should be able to interpret the remainder in context and realise that the school would require 9 coaches.

3 Students should be encouraged to record their working in a clear and systematic way to avoid little mistakes.

4 The numbers involved in the calculation provide a starting point for a discussion about efficient methods of calculation. When multiplying 29 by 20, for example, some students might instinctively use a formal method whilst others may double 29 and then multiply the answer by 10.

6 Students should be encouraged to record their working in a clear and organised way to ensure that they can easily state the answer that the question is looking for.

What do you think? 6.3A

These three questions provide opportunities for students to think more deeply about solving problems involving integers. Question 1 focuses on different methods for performing calculations. Question 2 draws students' attention to specific words in questions and how these are a clue as to what calculations should be performed. Question 3 is a more in-depth problem-solving question.

Practice 6.3B

1 addresses a common mistake of misaligning the numbers when calculating with decimals.

3 A common misconception here is to think that the answer will be 10 times smaller and therefore be 66.3, but students should realise that because each number in the calculation has been divided by 10 the answer needs to be divided by 100.

6 Some students may try to use a formal method for each calculation rather than related facts. They should think about what has happened to each number in the calculation and decide how this affects the answer.

7 Students might give an answer of £13.30 for part a because they have just added together the two values in the question. A bar model can help to explain why this isn't correct.

8 Some students might be tempted to divide by 5 and then multiply by 40 but should realise that they could just multiply by 8 instead.

What do you think? 6.3B

These questions provide opportunities for students to think deeply about calculating with decimals. In question 3, they could be encouraged to explore a variety of methods and discuss the pros and cons of each.

Practice 6.3C

1 to **4** provide opportunities for students to consolidate working with directed numbers.

5 Students should use their knowledge of inverse operations to work out the missing values.

6 Ensure that students don't make vague generalisations, e.g. "two minuses make a plus". Instead, they should use correct mathematical language and explain that the product of two negatives is positive.

What do you think? 6.3C

These three questions enable students to think more deeply about calculating with directed numbers. In question 2, ensure that students don't go straight to the calculation but think about the sign of the answer first.

Consolidate – do you need more?

These seven questions provide further opportunities for students to solve problems involving integers, decimals and directed numbers.

Stretch – can you deepen your learning?

1 Drawing a simple diagram can help students to answer this question correctly.

3 Some students may incorrectly think that they can purchase fish fingers individually.

4 and **5** Bar models can be used for support and may help students to organise their thinking.

Reflect

Students are encouraged to reflect on their learning by explaining in their own words how the language used in a question can help them to decide which calculations they need to perform. Question 2 encourages students to think about the rules for calculating with directed numbers. Ensure that they use correct mathematical language to help to avoid any misconceptions later.

6.4 Fractions

Small steps

- Add and subtract fractions (R)
- Multiply and divide fractions (R)
- Solve problems with fractions

Rationale

This chapter is designed to recap the four operations with fractions before moving on to solving problems. The first part of the chapter focuses on basic addition and subtraction of fractions whilst the second part focuses on basic multiplication and division. The last part combines the skills covered in Practice 6.4A and Practice 6.4B, and requires students to solve problems involving fractions. Although students will have performed many calculations involving fractions in previous years, this often proves a challenging topic. Concrete and pictorial representations can be used for support if needed.

Things to look out for

- Some students may just add/subtract the numerators and denominators without first finding a common denominator. Unitising can help to undo this misconception: e.g. if 3 eggs + 2 eggs = 5 eggs, then 3 sevenths + 2 sevenths = 5 sevenths. Pictorial representations such as bar models can also be useful.
- Students may find a common denominator when multiplying or dividing fractions and should realise that although this doesn't give an incorrect answer it overcomplicates the question as the answer will need to be simplified.
- Students often make mistakes when converting between improper fractions and mixed numbers because they're unsure of what each part represents. Bar models can be used for support.

Interleaving opportunities

- Knowledge of factors and multiples will be used when working with equivalent fractions.
- Comparison with collecting like terms can be used to help students to understanding why fractions need to have the same denominator before adding/subtracting them. They could think of different denominators as being unlike terms.
- The chapter provides opportunities to revisit algebra. Students could be asked to solve equations with fractional answers or substitute fractions into expressions.
- Fractions can be interleaved with geometry by asking students to find the area and perimeter of shapes for which the dimensions are given as fractions or mixed numbers.

Models and representations

- Bar models can be used for support throughout the chapter.

Behind the questions

Are you ready?

These five questions cover the essential knowledge for this chapter. Question 1 covers simplifying fractions. Students could be encouraged to compare their methods to see whether it is better to use the highest common factor or just any factor when simplifying. Questions 2, 3 and 4 require students to convert between improper fractions and mixed numbers. In questions 3 and 4, they may find it useful to draw bar models similar to the one in question 2. Question 5 revises adding and subtracting fractions with the same denominator. This is a good opportunity to remind students that two or more fractions must have the same denominator before they can be added or subtracted. In part c, students should be able to explain why $\frac{6}{6}$ is equivalent to 1 whole.

Practice 6.4A

1 Students should recognise that some calculations require both fractions to be adjusted whilst others require only one fraction to be changed. They should be encouraged to explain why this is the case.

2 allows students to consolidate their understanding of adding and subtracting fractions.

3 The visual representation is provided here for support. Students could be stretched to consider how the bar models also show that the answer is $\frac{25}{6}$.

4 and **5** provide practice at calculating with mixed numbers. Students could be encouraged to compare their methods with a partner's and discuss the pros and cons of each.

What do you think? 6.4A

These two questions provide opportunities for students to think more deeply about addition and subtraction of fractions. Question 1 invites them to consider different methods and when each is more appropriate, whilst question 2 addresses the common mistake that students make, of always multiplying the denominator.

Practice 6.4B

1 In part a, the fraction $\frac{1}{5}$ is used throughout and in part b the same integer is used. Students should be encouraged to think about how this affects the answer.

2 is designed to make students think explicitly about the area model for multiplication.

3 From part e onwards, the calculations include improper fractions and mixed numbers. As an extension, students could be asked to consider how simplifying first could make each calculation more efficient.

6 Students should be encouraged to use correct mathematical language when describing their process for dividing fractions. They should know that dividing by a fraction is the same as multiplying by its reciprocal.

What do you think? 6.4B

In question 1, students could be encouraged to use algebraic notation to support their reasoning for each part. Question 2 highlights the importance of being able to simplify before calculating, and students could be encouraged to break down the calculation into factors, for example $\frac{121}{180} \times \frac{36}{55} = \frac{11 \times 11}{36 \times 5} \times \frac{36}{11 \times 5}$, to make this easier.

Practice 6.4C

1 In part b, some students may add together only the two given sides and not include the other lengths.

4 A common mistake is to forget to divide by 2 when finding the area of a triangle.

5 Some students might prefer to write each measurement in millilitres before finding the answer.

7 Students can use complements to one whole to help them to answer this question. They should be encouraged to explore a range of different methods.

What do you think? 6.4C

These three questions provide further opportunities to solve problems involving fractions. Students should be encouraged to write out each step of their thinking for each question and to perform one step a time.

Consolidate – do you need more?

These questions provide further opportunities for students to practise solving problems involving fractions.

Stretch – can you deepen your learning?

1 revisits fractional and decimal equivalence.

2 Ensure that students use the correct order of operations.

5 reinforces the idea of simplifying before calculating, as seen earlier in the chapter.

6 to **8** look at fractions in algebraic contexts. In question 7, students could be given a pair of coordinates, e.g. (2, *b*), and asked to work out the value of *b*, and then be given (*c*, 2) and asked to work out the value of *c* to guide them towards the steps needed to find the expression for *a* in terms of *p*.

Reflect

Students are encouraged to reflect on their learning by explaining in their own words how to perform calculations involving fractions.

6.5 Standard form

Small steps

- Numbers in standard form (R)

Rationale

This chapter is designed to recap standard form, which was covered in detail in Book 2. Practice 6.5A starts by giving students the chance to practise converting between ordinary numbers and standard form. It's important that they understand the steps they need to take and can explain what each part of the standard form representation means. The exercise then moves on to calculating with numbers in standard form to allow students to consolidate their understanding. This is a good opportunity to revisit entering numbers in standard form on a calculator. Calculators can be used to check answers throughout the chapter.

Things to look out for

- Students may simply count the zeros when writing ordinary numbers in standard/ordinary form, e.g. they may think that $5.4 \times 10^6 = 54\,000\,000$. However, if they understand that $10^6 = 1$ million, then they should be able to explain why 5.4×1 million is not equal to 54 million.
- Students often make mistakes when calculations involve indices. Encourage them to write out their working in full to prevent this.
- Students often give answers that are not in correct standard form. Ensure that they are able to recognise when their answers are not correctly written.

Interleaving opportunities

- Knowledge of place value can be interleaved throughout to support understanding of standard form.
- Algebraic notation can be revisited and can be used to support understanding of the conditions for a number to be written in correct standard form, e.g. that $a \times 10^n$ is the correct format, where $1 \leqslant n < 10$ and n is an integer.
- Students can revisit substitution if the values being substituted are given in standard form.
- Laws of indices from Book 2 are revisited when calculating with standard form.

Models and representations

- Place value charts are useful for comparing large or small numbers.
- A calculator can be used to work with numbers in standard form.

Behind the questions

Are you ready?

These four questions ensure that students have the essential prerequisite knowledge for this chapter. Questions 1 and 2 focus on basic calculations with both integers and decimals. Question 3 revisits inequalities to ensure that students can correctly identify numbers that do and do not satisfy a given inequality. Question 4 revisits powers of 10: knowledge of these is essential for converting between ordinary numbers and standard form.

Practice 6.5A

1 Notice that each number has 3 in the hundred millions column and that the digits in the other place value columns change in the question parts. You could point out the similarities and differences between these numbers, and how these affect the answers when they are written in standard form.

2 is similar to question 1, but this time the focus is on numbers less than 1. The digit 8 remains constant in all parts but, in this question, its place value changes.

3 and **4** provide opportunities for students to practise writing numbers in standard form. Links should be made between the place value of the first significant figure and the power of 10. In parts g and h of question 4, watch out for students incorrectly reading the words as hundreds and thousands.

5 highlights the sorts of real-life numbers for which standard form is useful.

6 requires students to compare numbers in standard form. In part a, they should recognise that, because the power of 10 is the same, they can just compare the number parts. In part b, students may need to convert one of the numbers to be able to compare them. In parts c and d, students should recognise that they can compare the numbers by looking at the powers of 10.

7 Students often give answers that are not in correct standard form, and this question addresses this error. Encourage students to explain why the numbers are not written in correct standard form before correcting them.

8 provides practice at adding and subtracting numbers written in standard form. Some students may choose to write each number in ordinary form before performing the calculation and then convert back, whilst others may choose to adjust one of the numbers to make the powers of 10 the same. Students should be encouraged to compare these methods and discuss which is more efficient and when. Students should carry through their skills from question 7 to ensure that their answers are given in correct standard form for each part.

9 and **10** are similar to question 8 but here the focus is on multiplication and division of numbers written in standard form. Again, students should be encouraged to compare their methods and ensure that their answers are given in correct standard form.

11 to **13** provide opportunities for students to work with standard form in different contexts. These contexts can be used to reiterate the advantages of the use of standard form as a way of recording and comparing very big and very small numbers.

What do you think? 6.5A

These three questions require students to think in more detail about standard form. Question 1 can surprise students, and they often think that the answer can't be correct. In question 2, watch out for students trying to use the speed = $\frac{\text{distance}}{\text{time}}$ formula by rote rather than thinking carefully about the information given and what they're trying to find out. In question 3, some students will convert to ordinary form whilst others will convert the population of Northern Ireland to standard form. They should be prompted to discuss which method is more efficient and why.

Consolidate – do you need more?

These seven questions provide further opportunities for students to consolidate working with numbers in standard form. The focus should be on ensuring that all answers given are in correct standard form where appropriate.

Stretch – can you deepen your learning?

1 Students could be encouraged to recognise that the "$\times 10^8$" can be cancelled when written as a fraction, which can help them when calculating the percentage.

2 Watch out for students using an incorrect order of operations.

4 Students might need to be reminded of what is meant by the square root of a number.

5 Students should think carefully about what happens to the number when adjusting the power and then apply their understanding in an algebraic context.

6 Students should recognise that the inequalities given determine whether or not the answer will be in correct standard form.

7 introduces algebraic notation in the context of standard form.

8 interleaves various topics from earlier in the curriculum. Students could be encouraged to draw diagrams to support their thinking.

Reflect

Students are encouraged to reflect on their learning and explain in their own words how to write a number in standard form.

7.1 Percentage basics

Small steps

- Use the equivalence of fractions, decimals, and percentages (R)
- Calculate percentage increase and decrease (R)
- Express a change as a percentage (R)

Rationale

This chapter reminds students of the key conversions and calculations covered in Books 1 and 2. The images and representations should all be familiar. Students are encouraged to fluently convert between different forms of numbers and to recognise when calculator use would be appropriate. Students then revisit percentage increase and decrease and, once again, the multiplier approach is emphasised in preparation for its application to reverse percentage problems in Chapter 7.2. In the final section of the chapter, students consolidate prior work on expressing change as a percentage, using both the formula percentage change = $\frac{\text{change}}{\text{original}} \times 100\%$ and by setting up and solving equations to identify and interpret the multiplier used.

Calculations both with and without a calculator are explored, with the emphasis placed on making appropriate choices. Many of the answers will need rounding: students will have covered this in Block 12 of Book 2, so this provides an opportunity to consolidate this work in context.

Things to look out for

- Thinking that 7% = 0.7, for example. Discuss the meaning of % and link it to division by 100.
- Using non-calculator methods when calculator methods are more appropriate. Discuss which approach to take and when.
- Difficulties in working with percentages over 100. Use number lines to illustrate that the same relationships hold as for percentages below 100.
- Student relying on non-calculator methods. Encourage the use of and demonstrate the efficiency of multipliers.
- Difficulty in using the percentage change formula (e.g. using the wrong denominator).

Interleaving opportunities

- This chapter develops the learning of Block 10 in Book 2.
- There is ample opportunity to revise rounding during this block.
- Questions in this chapter include topics such as area, ratio, probability and use of inequality notation.
- The rest of Block 7 builds on this chapter.

Models and representations

- The bar model is used to illustrate fractions of amounts and the "build up from 10%" method.
- Number lines are very useful for converting between all three forms. These should be extended beyond 1 or 100% as appropriate.
- Fraction walls can support students when converting between related fractions, such as fifths and tenths.

Behind the questions

Are you ready?

The first two questions assess students' ability to convert between fractions, percentages and decimals in preparation for more challenging questions in Practice 7.1A. Question 3 offers a chance to review rounding, whilst questions 4 and 5 check students' understanding of finding simple percentages and fractions of amounts.

Practice 7.1A

1 The total number of squares in each grid is a factor of 100, which makes this a good starter question for revision. However, you may wish to explore with students the effects of changing the number of squares.

2 to **6** allow ample opportunity to convert between fractions, decimals and percentages. Students can explore misconceptions in question 4 and practise using inequality notation in question 6.

7 reminds students of the advantages of using bar models to answer these types of questions. You may wish to add more examples or ask students to make up their own.

8 gives students the opportunity to problem-solve and review their knowledge of probability at the same time.

What do you think? 7.1A

1 Encourage use of bar models. It is also worth exploring different approaches: for example, is it more efficient to convert to percentages first or to carry out the fraction calculation and convert later?

2 and **3** allow students to apply their skills to multi-step problems. Bar models could be used in both questions.

4 provides a chance to review area work and apply percentages in a different context.

5 is an opportunity to review algebra skills.

Practice 7.1B

1 and **2** have both been scaffolded into smaller steps and have pre-drawn bar models. This should help students to approach the remaining questions in this exercise.

3 to **5** test students' knowledge of multipliers since this underpins much of the forthcoming work.

6 and **7** have less scaffolding, so students are encouraged to develop their own strategies and to determine whether or not it would be appropriate to use a calculator.

8 addresses the common misconception that percentages are additive. Pictorial representations can demonstrate why this is not the case.

9 interleaves work on area.

10 is an opportunity to apply knowledge to a multi-step problem set in a real-life context.

What do you think? 7.1B

1 and **2** explore the effect of repeated percentage change. This will be looked at in more depth in Chapter 7.4.

3 to **5** allow students to investigate the equivalence of different calculations and why this might be the case.

Practice 7.1C

1 and **2** are scaffolded into smaller steps and allow students to build up strategies for future use.

3 to **5** involve expressing a change as a percentage. Scaffolded parts from questions 1 and 2 can be used to support students.

6 is a fluency task to allow students to make connections between multipliers and percentage change.

7 could work well as a discussion task about the cost of living and wages.

8 could provide another discussion topic: for example, "What does 'most improvement' mean?"

What do you think? 7.1C

1 and **2** provide extended problems where students need to read and interpret the text with care.

3 deals with the common misconception that to undo an x% increase, you would decrease the new amount by x%. This will be addressed in more detail in the next chapter.

Consolidate – do you need more?

This fluency practice could be used for homework or could be revisited for reinforcement at a later date.

Stretch – can you deepen your learning?

1 is an exploration of combined percentage change, using area as a familiar context.

2 allows students to explore the effect and non-linear nature of repeated percentage change.

Reflect

This encourages students to think about how and why their methods work as well as common misconceptions.

7.2 Reverse percentages

Small steps

- Solve reverse percentages with and without a calculator

Rationale

Students first met reverse percentages in Chapter 10.5 of Book 2, by way of a Higher step that not all students may have accessed. This chapter assumes no prior knowledge and looks at both calculator and non-calculator methods, with a focus on the efficiency of the multiplier method.

When using non-calculator methods, students are encouraged to draw bar models as pictorial representations to help them to determine the correct calculation when given x% of a number. When using a calculator-based approach, students are encouraged to set up and solve equations by identifying the relevant multiplier.

Time should be spent looking at key words and phrases that commonly indicate a reverse percentage question. This will support students in the next chapter when looking at mixed percentage problems.

Things to look out for	Interleaving opportunities
• Mixing up multiplication and division and/or using the wrong multiplier. Students need to be encouraged to ask themselves, "What percentage do I know?", "What do I need to find out?" and "What steps can connect these?" • Treating all problems as reverse percentage problems. • The common misconception that to undo an x% decrease, you increase the current amount by x%. Bar models can help to demonstrate why this isn't the case.	• Familiarity with the connections between fractions and percentages from Chapter 7.1 is assumed. • The multiplier method was introduced in Chapter 7.1. • Forming and solving simple one-step equations is a key technique. • Students will need to round answers appropriately. • Other topics, such as time, area and ratio, are interleaved within the practice questions.

Models and representations

- A bar model with "before" and "after" clearly visible is a good way of representing this type of problem. Careful labelling of the bar model should ensure that the right calculations follow.

Behind the questions

Are you ready?

Understanding and use of decimal multipliers is key to the algebraic approach of solving reverse percentage problems, so this is the focus of the first two questions. Question 3 provides a chance to assess students' ability to solve one-step equations.

Practice 7.2A

1 to **6** are designed to be accessible without a calculator. These questions could be used to compare and contrast calculator and non-calculator methods.

1 and **2** encourage proficient use of bar models and finding different parts of a whole when given one specific part.

3 and **4** extend the ideas met in the earlier questions to different contexts where the whole needs to be found when you are given a part. Students could be encouraged to draw their own bar models.

5 to **10** offer a variety of contexts for reverse percentage questions and support students in recognising key words and phrases that identify these types of problems, for example, "after an increase", "the original amount" and "before VAT was added".

11 addresses a common misconception and aims to show that 15% of the original amount is not the same as 15% of the final amount.

12 shows the connections between percentage increase/decrease problems, expressing change as a percentage and reverse percentage problems. Each time, there are three parts (original amount, percentage change and final amount) and one is unknown. Students could be asked to compare and contrast their methods depending on which of the three parts they are trying to find.

13 to **15** offer further opportunities to practise newly acquired skills in a variety of contexts. These also provide students with the chance to look for key words and phrases that help to determine the nature of the questions. Student should always round their answers appropriately, for example to 2 decimal places for money.

What do you think? 7.2A

1 and **2** extend the concept of reverse percentages to repeated change, so that students need to undo two separate percentage changes. This is a Higher step and will be explored further in Chapter 7.4.

3 offers a chance to address misconceptions. A starting value isn't given, so students can use different values and see what they notice before trying to generalise. They could also investigate whether the same would hold true for different percentage changes.

Consolidate – do you need more?

This provides ample opportunities to develop fluency and problem-solving skills with a variety of questions. The first four questions are designed to be accessible without a calculator.

Stretch – can you deepen your learning?

1 is a multi-step problem in which students first need to use the given information to identify the percentage change (thus reviewing Chapter 7.1), before going on to finding the original amount.

2 is an opportunity for students to think about the equivalence of finding 80% and reducing by 20% and could form the starting point for a discussion.

3 interleaves work on area and explores ideas around scale factors of enlargement.

4 reminds students of the importance of context. In this example, there cannot be a fractional number of people.

5 and **6** provide extended opportunities to look at much more complex questions that draw together work on percentages from this chapter and the previous one, as well as work on ratio from Book 2.

Reflect

These questions focus on recognising when a problem is a reverse percentage problem as well as on how to tackle such a question. Students are encouraged to compare and contrast different approaches.

7.3 Solving percentage problems

Small steps

- Recognise and solve percentage problems (non-calculator)
- Recognise and solve percentage problems (calculator)

Rationale

The main purpose of this chapter is to give students a chance to explore a mixture of percentage problems so that they can identify problem types and choose the correct approaches.

Students can often struggle to determine which type of problem they are facing, which methods would work, and whether or not it would be sensible to use a calculator. The chapter is split into two parts: the first focuses on non-calculator approaches and the second develops efficient calculator use.

Problem-solving like this is a challenge, and ample time should be given to this chapter. Students will get the opportunity to recall and consolidate work from Books 1 and 2.

Things to look out for	Interleaving opportunities
• Difficulty in choosing the correct approach to take with a question, e.g. multiplication or division, or whether or not it is a reverse percentage question. Students could start by looking through the practice questions to identify types and methods before carrying out the calculations. • Difficulty in understanding or mixing up key vocabulary such as "profit", "loss", "gain" and "depreciate". Reviewing key vocabulary from the previous books would be worthwhile.	• Work from Block 10 of Book 2 is reviewed. • All the work from Chapters 7.1 and 7.2 will also be reviewed in this chapter. • Students will have the opportunity to practise rounding to a sensible degree of accuracy throughout the calculator section. • You can set problems about percentages in almost all other areas of the mathematics curriculum, especially number-based or number-heavy topics.

Models and representations

- Students are encouraged to use bar models for representing problems, as this will help them to decide what approach to take. Ask questions about the model to support students, e.g. "Are you looking within the bar, or are you changing the size of the bar?"
- The formula from Chapter 7.1, percentage change = $\frac{\text{change}}{\text{original}} \times 100\%$, is a useful abstract representation for support when expressing a change as a percentage.

Behind the questions

Are you ready?

1 addresses rounding, as this will be a key skill in the calculator part of the chapter.

2 to **5** review all the key skills related to percentages that students have met so far. These questions could be used to indicate whether prior chapters from this block or from Block 10 of Book 2, need to be revisited.

Practice 7.3A

1 and **2** review key skills whilst reminding students of the usefulness of pictorial representations.

3 and **4** are applications of percentage increase and decrease.

5 revises work from Book 2 on expressing one value as a percentage of another.

6 and **7** use part–whole comparisons. In question 6, 84 represents 70%, and in question 7, 63 represents 15%. Both questions can be modelled using bar models, which will help students to consider the best approach.

8 is a reminder of reverse percentage questions.

9 to **11** offer a range of different contexts in which students need to apply one or more of the above skills to solve multi-step problems.

What do you think? 7.3A

1 may encourage discussion of what is meant by "better value for money" before exploring the main question. No values are given, so students could pick their own carton size and price and then consider whether changing these would change the answer to the question.

2 This question and others like it often confuse students as no starting value has been given. Students could choose a starting value or they could consider the problem from a purely algebraic perspective.

Practice 7.3B

1 and **2** are simple percentage increase/decrease questions, with a reminder about the usefulness of bar models to support students' thinking and approaches.

3 revises expressing one number as a percentage of another in a simple context.

4 addresses the concepts of profit and loss, which were met in Chapter 7.1.

5 is a one-step reverse percentage question that will revise work from Chapter 7.2.

6 to **8** provide contexts in which students can apply their knowledge; question 7 interleaves work on the volume of cuboids and converting between metric units of measurement.

9 allows a chance to explore equivalent calculations. Students could be asked to think about *why* some of these are equivalent.

10 to **12** provide further examples of multi-step problems set in context that require interpretation.

What do you think? 7.3B

1 makes a key point about the importance of language and identifying which part of the possible information has been given. Here, £2.30 represents the 10% increase.

2 and **3** offer extensive, multi-step problem-solving opportunities. They include work on ratio.

Consolidate – do you need more?

This exercise provides further mixed questions on percentages to help students to develop fluency and problem-solving skills.

Stretch – can you deepen your learning?

1 to **4** offer more challenging multi-step problems in a variety of contexts and consolidate work on percentages, fractions and ratios.

5 is a more complex reverse percentage question.

Reflect

Students could produce a list of key words or an example of each type of question for future use.

7.4 Repeated percentage change (H)

Small steps

- Solve problems with repeated percentage change (H)

Rationale

This chapter is a Higher step and students should only embark on it if they have developed good fluency with the work from earlier chapters in this block.

The concept of repeated percentage change in the context of compound interest will be met in the next block. This chapter focuses on a variety of contexts and is not limited to the same percentage change being applied.

Although it is possible to access and answer the questions by applying one change at a time, the formula for repeated percentage change is introduced as a shortcut approach. This will be used extensively in the next block, so its application is not essential at this stage.

This chapter also serves to reinforce key ideas such as repeated change being non-linear and non-additive.

Things to look out for	Interleaving opportunities
• Thinking that a 10% reduction followed by a 10% reduction is the same as a 20% reduction. • Misunderstanding the language used, e.g. "depreciation", and therefore interpreting questions incorrectly. • Thinking that a 25% reduction applied four times means that there will be nothing left. Graphing some of the examples will show the non-linear nature of these problems.	• This chapter develops learning from Chapters 7.1 to 7.3; for example, the use of multipliers and reverse percentages. • There is ample opportunity to revise rounding during this block. • Many of the financial elements of the content will be reviewed in the next block.

Models and representations

- Bar models can be used to illustrate repeated change and to effectively show that since the whole changes after each percentage change, so will the part/percentage.
- Bar models can help students to identify the problem type and the correct approach to take.

Behind the questions

Are you ready?

The first three questions review work from Chapters 7.1 to 7.3 since this chapter relies on students' fluency with those skills. Question 4 is a quick chance to review rounding in preparation for this largely calculator-based chapter.

Practice 7.4A

1 would make a good discussion point. Students could explore whether this works with other starting amounts and other percentages. Some may be able to generalise their findings.

2 involves two percentage changes. This can be accessed without a calculator and may help students to clarify their thinking about repeated percentage changes.

3 to **5** provide different contexts for repeated change. Although questions 3 and 4 can be answered by applying each percentage change in turn, question 5c demonstrates the usefulness of the formula as a more efficient tactic.

6 and **7** provide some different contexts. Question 7 reminds students of the need to round sensibly based on the given context.

8 addresses a common misconception. Note that the starting value for the rent has not been given so students could try values of their own to see what would happen.

9 reinforces the formula for repeated percentage change.

10 offers a chance to explore repeated percentage change in the context of the volume of a cube and leaves students with an open-ended investigation in part d.

What do you think? 7.4A

1 provides another chance to address misconceptions and is another question in which a starting value is not given, so it can be open to exploration.

2 helps students to investigate the non-linear nature of repeated percentage change. Students could record their results for the height after each week and draw a graph showing the growth of the plant over time.

3 Bar models or chosen values could be used to clearly demonstrate Jakub's misconception.

4 This question could be approached using trial and error, by picking different values to increase and decrease, or by considering an algebraic approach using multipliers.

Consolidate – do you need more?

This exercise provides further questions and contexts for applications of repeated percentage change. It can be used for homework or it could be revisited at a later date.

Stretch – can you deepen your learning?

1 and **2** provide a chance to review increasing/decreasing by a fraction as well as considering percentage changes.

3 and **4** explore repeated percentage change from the perspective of combining multipliers: for example, a 20% increase followed by a 20% increase can be represented by 1.2 × 1.2 = 1.44, which represents a 44% increase. A follow-up question could be to ask why this is bigger than 20% + 20%.

5 provides another chance to explore repeated change through trial and improvement. Again, graphs could be drawn to consider the non-linear nature of the changes.

Reflect

1 encourages students to consider the effect of combining multipliers.

2 encourages students to think about why the formula is a useful tool for efficient calculation.

8.1 Interpreting bills and statements

Small steps

- Solve problems with bills and bank statements

Rationale

In this chapter, students will develop their understanding of bills, receipts and bank statements.

Many students are not aware of how utility bills are calculated using meter readings and cost per unit of gas or electricity, so in this chapter they will have the opportunity to learn more about these types of bills whilst consolidating numeracy work and prior work on percentages.

Students may benefit from looking at mock/old bank statements so that they can become familiar with the language of banking.

Things to look out for	Interleaving opportunities
• Many students will not be familiar with the language of financial mathematics. A class glossary of key words can be created as new words and phrases are met. • Students may not be familiar with the concept of overdrafts and negative bank balances, so clarification may be needed. • Many students will not have seen a gas or electricity meter so it may be helpful to show some pictures.	• Students will review their work on financial mathematics from previous books. • There are opportunities to include work on percentages. • Work on negative numbers can be reviewed through the context of banking, i.e. debits and overdrafts.

Models and representations

- Students may benefit from looking at mock/copied documents such as bills, bank statements and receipts so they can become familiar with the elements that make up such paperwork.

Behind the questions

Are you ready?

These questions provide a quick chance to review key numeracy skills such as working with money, rounding to 2 decimal places and working with negative numbers.

Practice 8.1A

1 enables students to analyse a receipt and its key features whilst practising their numeracy skills. This can be done with or without a calculator.

2 and **3** develop students' problem-solving skills in the context of shopping. Question 3 offers a chance to review units of measurement and unit conversions.

4 and **5** allow practice at calculating utility bills using meter readings, and both require students to convert between pounds and pence. Costs per unit of gas or electricity are often given in pence per unit, and can be non-integer values. Some students may need help to convert 22.8p into pounds.

6 is a chance to address common errors when dealing with mixed units of money.

7 is a more complex bill problem, which involves two variables (phone calls and text messages). Students can be asked to compare older tariffs from the late 1990s with current tariffs as an investigation to see how much cheaper mobile phone use has become.

8 offers a reminder of the meaning of overdrafts and being overdrawn. These terms could be added to students' glossaries.

9 involves numeracy in the context of handling money and negative number calculations.

10 allows students to develop some familiarity with bank statements and the concepts of credit and debit whilst practising their numeracy skills.

What do you think? 8.1A

1 introduces the concept of bank charges, so students will need to add the fee to the overdrawn amount. The language used emphasises that overdrafts are essentially money *owed* to a bank.

2 is a more complex question set in a simple shopping scenario, in which students will need to work backwards to find the missing value. This is a good question to use to remind students of the need for clarity and logic when writing up solutions to multi-step problems.

3 is another working-backwards problem, this time involving an electricity bill. Students need to take care with the mixed units in the question.

Consolidate – do you need more?

This exercise provides further opportunities to practise working with money and solving problems in the context of shopping and paying bills. The questions can be used as review activities or set as future homework tasks.

Stretch – can you deepen your learning?

1 involves several different steps and helps students to refine their problem-solving skills. Electricity is charged at two different rates and there is an added administration fee (consolidating percentages) to take into account.

2 could be developed into an investigation involving graphing the costs of the two tariffs and considering the meaning of the point of intersection in the context of the question.

Reflect

This encourages students to investigate different types of bills and to discuss the different ways in which they may be calculated. Students could further investigate tariffs using money-comparison websites.

8.2 Interest

Small steps

- Calculate simple interest
- Calculate compound interest

Rationale

In this chapter, students will build on much of the work in Block 7.

The first section focuses on the concept of simple interest and the second picks up the Higher step work of Block 7 by looking at repeated percentage change in the context of compound interest. No prior knowledge is assumed and the concept is met intuitively rather than through a formula.

Although compound interest is beneficial in savings accounts, you should also discuss loans and the increasing costs incurred when compound interest is applied to them.

Things to look out for

- Students may mix up the two different types of interest. Bar models can be used to show the differences between the two visually.
- Students should be encouraged to read the questions carefully and look for key words to appreciate, for example, the difference between being asked for the interest accrued and the total savings after a given period of time.

Interleaving opportunities

- Students will review their work on financial mathematics from previous books.
- The work of Block 7 is reviewed, consolidated and extended by introducing repeated change more formally.
- There are opportunities to review trial and error as a method of problem solving.
- Simple algebraic equations can be used for some of the problem-solving questions.

Models and representations

- Bar models should be used to demonstrate percentages of amounts and percentage increase and decrease. They can also help to clarify reverse percentage questions by considering what the given amount represents.

Behind the questions

Are you ready?

Questions 1 to 4 review students' knowledge of percentage calculations, including the use of multipliers, which will be essential in this unit. Question 5 is an opportunity to review rounding, power calculations and the correct use of a calculator.

Practice 8.2A

1 and **2** offer scaffolded opportunities to find simple interest by rephrasing them as "percentage of" problems. Bar models can be used at this stage to support students if needed.

3 and **4** offer further practice at finding both interest and total balances in less scaffolded examples. Students can use the earlier questions to support their thinking.

5 and **6** present interest in the context of fees for purchase rather than as a savings bonus. As well as practising their mathematics, students will develop their understanding of real-life financial transactions.

7 is a nice opportunity to discuss the calculations behind the two examples and why they are mathematically identical. Writing the percentage as a fraction and looking at rules of fraction multiplication will help to clarify this for students.

What do you think? 8.2A

The three questions in this section demonstrate different levels of problem solving.

1 involves simply finding the number of years.

2 Students will need to consider how to use the annual gain in pounds to work out the annual percentage rate. This provides a chance to review expressing one number as a percentage of another.

3 has both the initial amount and the interest rate missing. Complex mathematics is not required to answer it, but it needs a logical approach. Students can find the amount of interest earned per year (in pounds) and subtract repeatedly to get the initial amount. At this stage, this problem is similar to question 2.

Practice 8.2B

1 and **2** offer step-by-step approaches to solving compound interest problems and also remind students of the need to distinguish between interest earned and total bank balance.

3 interleaves and contrasts with simple interest from the earlier part of the chapter.

4 shows the rapidly growing nature of compound interest. This can be explored and discussed further, for example by considering the impact of this on interest on loans and also by comparing it with simple interest over the same period of time. Graphing software could be used to demonstrate this visually.

5 offers a chance to compare the two calculations and to consider why they are not the same.

6 and **7** involve an element of trial and error, which is a valid problem-solving strategy in this sort of context.

8 reminds students of the formula for finding repeated percentage change.

What do you think? 8.2B

1 is a multi-step problem which allows students to consider debt in more depth and to appreciate how it can grow quickly. Students can be asked to consider how much Zach ends up paying back and how this compares with the amount of the original loan.

2 addresses a misconception around the multiplicative nature of compound interest. At a given interest rate, any amount of money would take the same time to double. Students can try to use their algebra skills to explore this further.

Consolidate – do you need more?

This exercise provides plenty of further practice in the skills learnt in the chapter, both for fluency and to develop problem solving. These questions can be used for homework or for revision.

Stretch – can you deepen your learning?

1 allows students to explore the way in which savings can grow. Graphical software can be used to compare the two scenarios.

2 looks at the length of time for an investment to triple where the initial amount is not given. Students can use different initial values to see that this doesn't actually matter.

3 to **5** mirror work on simple interest from earlier in the chapter. Question 3 involves finding a missing percentage, which can be done using trial and error or by setting up an equation that requires finding a cube root to solve. Question 4 requires the use of trial and error, and question 5 asks students to find the initial amount and the interest rate. This can be done by finding the multiplier from year 4 to year 5 and then working backwards.

Reflect

1 Students are asked to explain the difference between the two types of interest, using examples to illustrate it.

2 is a chance to reflect on and explore further the effect of interest rates on the cost of debt.

8.3 Taxes

Small steps

- Solve problems with value-added tax
- Calculate wages and taxes

Rationale

In this chapter, students look at value-added tax and review their percentage problem-solving skills. Although the focus is on the UK VAT rates of 5% and 20%, there are opportunities to look at rates of tax from other countries. Students could be encouraged to explore VAT rates across the globe as an additional investigation.

The second part of the chapter is concerned with wages and income tax. It deals with hourly rates of pay and overtime as well as annual salaries and the calculation of income tax according to income bands. Again, the focus is on the UK tax structure, but opportunities have been provided for comparison between different nations.

Things to look out for	**Interleaving opportunities**
• Students may need reminding about solving reverse percentage questions when finding prices before VAT was added. • The language of overtime, time and a half and double time may need to be emphasised. • Calculating income tax using tax bands can be complex and will need to be clearly demonstrated.	• Students will review their work on financial mathematics from previous books. • The work of Block 7 is reviewed and consolidated: finding and writing percentages, percentage increase and decrease, and reverse percentage problems are all included. • Students can revisit international comparisons in the next chapter once they have met and worked with exchange rates.

Models and representations

- Bar models should be used to demonstrate percentages of amounts and percentage increase and decrease. They can also help to clarify reverse percentage questions by considering what the given amount represents.

Behind the questions

Are you ready?

Questions in this exercise review basic numeracy, working with money and key concepts from Block 7, including the use of multipliers.

Practice 8.3A

1 is a scaffolded example to remind students about finding a percentage of an amount. Students should be encouraged to use the multiplier method with 5%, which is half of 10% and so should be fairly simple. You could discuss the real rate of VAT on most goods in the UK (20%).

2 to **4** give students a chance to practise adding on VAT. They can be done with or without a calculator.

5 This question offers a chance to compare the current rate of VAT with the previous rate of 17.5%. A nice discussion would be to consider different approaches. Most students will probably find the two prices including VAT and then find the difference, but you could show that finding 2.5% will be sufficient to answer the question.

6 adds a different context and an extra step to a VAT problem.

7 is a chance to review reverse percentage work from Block 7.

8 since many countries use the euro, this is a chance to compare VAT in two different countries without worrying about the exchange rate (which will be met in the next chapter).

9 is a multi-step problem designed to review this section and to develop students' problem-solving skills.

What do you think? 8.3A

1 offers a really good chance to explore facts and myths about percentages. Although it's true that the order in which the fees are applied doesn't matter (owing to the commutative nature of multiplication), this would not be the same as finding 25% of the original amount.

2 and **3** allow students to explore multi-step problems that include working in reverse. Once again, the need for clarity in written communication should be emphasised.

Practice 8.3B

1 and **2** require calculations involving hourly rates of pay.

2 introduces the concept of overtime as seen in the worked example.

3 and **4** allow students to solve problems that require working backwards. Question 4 involves finding the rate of pay given the overtime pay.

5 and **6** involve finding annual income tax. Both questions have been split into smaller steps to help students to understand how income tax is calculated.

7 is a more complex problem that allows students to consolidate the key concepts of this section as well as to review their previous work on percentages.

What do you think? 8.3B

1 involves working backwards to consider how many hours can be worked before paying income tax.

2 uses the currency of Italy and France as well as their taxation structures to compare and contrast salaries.

Consolidate – do you need more?

This exercise provides plenty of further practice on the concepts introduced in this chapter. These questions can be used for revision or as additional practice if needed.

Stretch – can you deepen your learning?

1 is similar to the questions from Practice 8.3B but there is no scaffolding offered. Students need to use the table of values to find the total income tax and net pay.

2 and **3** Although exchange rates will be formally met in the next chapter, these questions allow students to hone their problem-solving skills. There is a lot of information – exchange rates, VAT and, in the case of question 3, additional fees – to consider.

Reflect

1 can be turned into a really good budgeting project. Now that students understand more about the cost of living, they can begin to plan an annual or monthly budget. Local listings can be used to judge rental/mortgage costs.

2 is an opportunity for students to investigate other deductions, such as National Insurance. They could also explore pensions and student loan contributions.

8.4 Solving problems with money

Small steps

- Solve problems with exchange rates
- Solve unit pricing problems

Rationale

This chapter develops students' abilities to reason with money and solve increasingly complex problems. The first section explores exchange rates, allowing students to develop fluency when moving between currencies as well as solving more extensive import/export-type problems.

The second part of the chapter explores best buys and looks at different types of offers when shopping. The focus is on the unitary method and comparing the cost per item/per gram rather than, say, scaling both quantities up. This topic will be revisited in more depth in Block 13.

Things to look out for	Interleaving opportunities
• Students should be encouraged to develop a sense of the multiplicative nature of exchange rates and to explore different ways of moving between the two, e.g. using ratio. • Although cost per item is a more natural approach, the chapter also explores amount per unit cost. Students can easily mix up the two. Time should be taken to link the calculation with the resulting unit: amount ÷ price = amount per pound/penny.	• Students will review their work on financial mathematics from previous books. • When working with exchange rates, students can further explore graphical representations to consolidate work on linear graphs. • Students will explore alternative approaches to best buys in Block 13.

Models and representations

- Bar models should be used to demonstrate the unitary method so that students have a visual understanding of what cost per unit or amount per unit cost mean.
- Proportion diagrams can help to emphasise the multiplicative nature of exchange rates.
- Real-life offers (such as on supermarket websites) can be used to provide a more concrete anchor for best buys and offers.

Behind the questions

Are you ready?

Questions in this exercise review basic numeracy skills involving money as well as providing a quick review of equivalent ratios, which can be used when solving exchange rate problems.

Practice 8.4A

1 allows students to make some conversions which can help to highlight the multiplicative nature of the exchange rate, e.g. that doubling the number of pounds doubles the number of pesos.

2 provides ample opportunity to convert between British pounds and a variety of other currencies. Questions involve working in both directions.

3 and **4** offer typical questions involving different exchange rates.

5 to **7** involve making comparisons. Students need to be clear about which currency they are converting to in order to make the comparison. Questions around which currency would be best can be explored and discussed.

8 introduces commission as an additional concept and offers a chance to review work on percentages. Students could be asked, "Does it matter whether the commission is charged before or after the exchange?"

9 Students will look at these kinds of conversion graphs in more detail in Chapter 13.1, but this is a good chance to lay the foundations for direct proportion represented in graphical form.

What do you think? 8.4A

1 extends students by working in two currencies at the same time so that they will need to make a decision about the point at which they need to carry out a currency conversion.

2 gives students the opportunity to compare exchange rates. It is not immediately obvious which offers better value for converting to euros. An extension would be to consider which rate is better for the reverse conversion.

3 has several different concepts that need to be addressed: exchange rates, import taxes and other fees (in different currencies) plus a chance to review the concept of percentage profit.

Practice 8.4B

1 and **2** are both scaffolded and provide students with a blueprint for tackling other, similar, questions.

3 and **4** explore best buys from the perspective of best value for money. Students can be encouraged to discuss different approaches.

5 changes the approach slightly – 6 DVDs are wanted so the value for money of each website needs to be explored in this context. This makes it more like a real-life scenario.

6 is another question set in a context that focuses on a required quantity.

7 addresses the confusion around price per unit or amount per unit price. Students should be encouraged to show units when calculating to avoid confusion.

8 provides a chance to consider different approaches and explore different strategies. Some similar examples may already have been discussed when answering the earlier questions.

What do you think? 8.4B

1 is more complex than it first appears. Students need to do more than one calculation to work out the answer.

2 is another, real-life, example where the quantity required matters. You could discuss which cat food is best value in terms of price per unit and the total eventual cost of the purchase if more than the required amount ends up being bought.

3 is a challenge question that requires students to think about the range of possible values subject to the given constraints.

Consolidate – do you need more?

This provides opportunities to develop fluency and problem-solving skills related to exchange rates.

Stretch – can you deepen your learning?

1 challenges students to use two different exchange rates to come up with a third. There are lots of possible approaches, and this would make an excellent discussion task,

2 is a chance to practise best buy questions as well as unit conversions between litres and millilitres and the volume of a cuboid.

3 is a typical real-life choice question. Often, the choice of an item is not simply based on cost per unit: there are other factors to consider, such as delivery costs.

4 explores different types of offers often seen in supermarkets and other shops and asks students to explore, compare and contrast them and to consider the effect of changing the variables (price versus number of books).

Reflect

1 It's important to realise that real life doesn't necessitate buying the item that offers the cheapest unit cost. You would not purchase 3 kg of butter just because it's best value for money. Other factors, such as food wastage, sell-by dates, household consumption and storage, all affect choices.

2 is a chance for students to investigate the process of converting currencies and the fees they may be charged.

9.1 Angles review

Small steps

- Angles in parallel lines (R)

Rationale

Students begin by reviewing the angle facts they met in Books 1 and 2 as these will be the foundation of the geometry content of the rest of this block. However, the focus of this chapter is on formal justification and giving reasons for answers, and students should be encouraged to do this from the start.

It is intended that the questions can be answered with a calculator but you could use this chapter to review numeracy skills if you wish.

Things to look out for	Interleaving opportunities
• Incorrect language and incomplete reasoning (for example, saying "because it is a triangle"). Incomplete reasons should be challenged so that students appreciate the need for precise language. • Mixing up the rules and names for angles in parallel lines. • Not being able to identify which angles are corresponding, alternate or co-interior. • Sometimes students may not appreciate that the rules work both ways, e.g. that the knowledge that co-interior angles add up to 180° can be used to prove that two lines are parallel by showing that the two angles inside them along a transversal add up to 180°.	• Work on angles from Books 1 and 2 is reviewed here. There is no new content in this chapter so it provides the opportunity to consolidate. • Students can use the chapter to review non-calculator numeracy. • Three-letter angle notation can be revised. • Properties of shapes are revisited in the context of angles in triangles and quadrilaterals.

Models and representations

- Straws can be used to model the angle facts.
- Dynamic geometry software can be a quick way to recall and revise all the key facts and to see the angles changing as the transversal is moved.
- Students should be encouraged to draw their own diagrams so that they can add in extra labels, etc. to support their reasoning.

Behind the questions

Are you ready?

These questions provide a quick review of key numeracy and number bonds to 180 and 360. They also cover key facts related to angles and straight lines that students will have met in the earlier books.

Practice 9.1A

1 reminds students of the angles in parallel lines and the relative positions of the angles within the lines, and reviews their understanding of this.

2 offers practice at one-step and two-step problems involving angles in parallel lines. Students are asked to give reasons for their answers to encourage them to get into good habits from the start.

3 and **4** review angles in triangles and quadrilaterals in increasingly complex problems. Students could be encouraged to sketch the shapes and to label all the angles as they work them out rather than holding this information in their heads.

5 is a chance to address a common misconception that all angles along a straight line (rather than adjacent angles on a straight line) add up to 180°.

6 and **7** apply the reverse of the facts about angles on parallel lines by asking students to determine whether the lines are parallel based on the given angles. Allow sufficient time to explore these questions, and ensure that students appreciate that the rules for relationships between angles can be used to prove geometrical facts.

8 brings together angles in parallel lines and some common quadrilaterals. Students can become accustomed to the visual representation of angles in parallel lines as being a pair of parallel lines with a transversal. It's important that they can recognise these angle relationships in more complex shapes and are able to apply their knowledge of angle facts.

9 extends question 8. If one angle in a parallelogram is given, the remainder can be found using facts about angles on parallel lines.

What do you think? 9.1A

1 is a more complex take on angles in parallel lines and incorporates two pairs of parallel lines. Students now need to realise that the angle facts apply across both set of lines.

2 extends the challenge of identifying parallel lines. Here there are four lines and students need to work out which are parallel. A copy of the diagram or sketch will be helpful. Students can be asked to make up their own and to consider the minimum amount of information needed to be able to answer the question.

3 introduces algebraic notation and methods. This will be met in more detail in Chapter 9.3 but, for the higher attaining students, this may be a good opportunity to explore how their algebra skills can help them to solve more complex angle problems.

Consolidate – do you need more?

This section provides further opportunities to practise applying key angle facts in a variety of contexts. The focus is on reasoning and giving correct and complete reasons. These tasks can be used for homework or revision.

Stretch – can you deepen your learning?

1 involves using angle facts in slightly more complex examples. As well as parallel lines, students need to consider the hatch marks and what these tell them about the properties of the shapes involved.

2 is a good question to remind students that some of the information that they are given isn't always visible. Here, the 90° angles are not marked on to the squares, which makes the problem seem much more complex than it is. Again, sketching the shapes and adding in this extra information will be useful.

3 is similar to question 2 and encourages problem solving at a deeper level.

4 is another chance to review algebra skills in solving equations, this time with unknowns on both sides. This will be looked at in more detail in Chapter 9.3, but some students will be ready to try it at this stage.

Reflect

This gives students a chance to review all the angle facts they know and to create notes/revision cards or posters which will prove extremely helpful as they work through the remainder of the block.

9.2 Chains of reasoning

Small steps

- Solve angle problems using chains of reasoning

Rationale

Here, students will extend the work of Chapter 9.1 by looking at more complex problems and those which require several steps to solve. Students often struggle with these as they don't always appreciate that other angles can be found, and instead focus on the unknown that they can't immediately work out.

Once again, students should be encouraged to give reasons for each step of their working when solving more complex problems, using three-letter angle notation if appropriate. When this isn't possible, students need to develop other strategies, such as labelling some of the unknown angles themselves.

Things to look out for	**Interleaving opportunities**
• Incorrect language and incomplete reasoning (for example, saying "because it is a triangle"). Incomplete reasons should be challenged so that students appreciate the need for precise language. • Mixing up the rules and names for angles in parallel lines. • Not being able to identify which angles are corresponding, alternate or co-interior. • Sometimes students may not appreciate that the rules work both ways, e.g. that the knowledge that co-interior angles add up to 180° can be used to prove that two lines are parallel by showing that the two angles inside them along a transversal add up to 180°.	• Work on angles from Books 1 and 2 is reviewed here. • Three-letter angle notation is revised. • Properties of shapes are revisited in the context of angles in triangles and quadrilaterals. • There are some opportunities for extension work using algebraic notation, although this will be looked at in more detail in Chapter 9.3.

Models and representations

- Straws can be used to model the angle facts.
- Dynamic geometry software can be a quick way to recall and revise all the key facts and to see the angles changing as the transversal is moved.
- Students should be encouraged to draw their own diagrams so that they can add in extra labels, etc. to support their reasoning.

Behind the questions

Are you ready?

These questions provide a quick review of key numeracy and number bonds to 180 and 360. They also provide a chance to review key angle facts and correct reasoning statements related to angles in parallel lines.

Practice 9.2A

1 begins with less complex problems and a review of using three-angle notation. This question could be used to model how answers should be set out with reasons clearly given.

2 provides plenty of examples for further practice. Students may benefit from sketching the diagrams so that they can label the angles that they find in the intermediate steps. There are different ways to find each missing angle, and students can work in pairs to compare their strategies.

3 involves finding missing angles in parallelograms, and students need to use angle facts about parallel lines and triangles to find them.

4 addresses common errors. Students often struggle to identify which angles are alternate, corresponding and co-interior. Highlighters and coloured pencils can help with the visualising. It can be difficult to spot Ed's error, so students may wish to ignore his working and solve the problem themselves first, before making comparisons with Ed's method.

5 at first looks complex but it's a chance to remind students about angles on straight lines and vertically opposite angles and the relationship between the two angle facts.

6 and **7** provide examples of questions that involve reasoning. Students should be reminded of the need to set out their work in a logical order and to give reasons for each stage.

8 reviews the little-used alternate exterior angles and the fact that they are equal. Students are asked to consider the definition of alternate angles (here they are both exterior to the parallel lines) and to verify for themselves that they are, indeed, equal.

What do you think? 9.2A

1 and **2** both involve overlapping triangles. Students have more than enough information to go on but will always think they don't! The angles in the equilateral triangles are not marked on, so students should be encouraged to add those in for themselves. To reinforce the idea that the focus should be on what can be found out (rather than an end goal), question 1 explicitly asks students to find *all* possible angles in the diagram.

3 offers a chance for an engaging investigation involving strips of paper, which are readily available. Using the paper and making the shapes will help students to think about the overlapping parts and the connections between angles in parallel lines and the paper strip.

Consolidate – do you need more?

This section provides further opportunities to practise applying key angle facts in increasingly complex shapes. As well as angles in parallel lines, students review their knowledge of different types of quadrilaterals and triangles. These questions can supplement the earlier exercise or be used for homework or revision at a later date.

Stretch – can you deepen your learning?

1 is almost a question involving proof. Students use angle facts to show that a triangle is equilateral. A discussion around proofs, "show that" and demonstration-type questions may be useful at this stage.

2 is a chance for students to extend their skills by applying a little algebraic knowledge to solving angle problems.

3 is investigational in nature and, because it does not include any diagrams, students are free to explore the question in their own way.

Reflect

1 reminds students that methods are not reasons and that reasons must give the mathematical rule or relationship being used.

2 Students could be asked about the minimum amount of information needed to create a problem that can be fully solved.

9.3 Angles and algebra

Small steps

- Solve angle problems with algebra

Rationale

In this chapter, students will add another layer of complexity to the work of Block 9. Algebraic notation is reviewed and students are required to solve equations in order to find missing angles and variables. Some students may struggle with moving from the angle facts into equations, so time should be spent making connections between the language of angle facts and that of algebraic notation.

Bar models can be helpful here for setting up the equations and for helping students to model their solutions.

Things to look out for	Interleaving opportunities
• Some students may find it difficult to transfer their words (vertically opposite angles are equal) into an equation, so this may need to be modelled to begin with. • Mixing up rules can lead to incorrect equations being set up. • Some of the questions in this chapter have multiple steps: identifying the angle fact, setting up and simplifying an equation, and then solving it. Students should be reminded of the need for clarity in their written solutions.	• Work on angles from Books 1 and 2 is reviewed, including angles in polygons. • Three-letter angle notation is revised. • Properties of shapes are revisited in the context of angles in triangles and quadrilaterals. • Simplifying by collecting like terms is used throughout. • Solving equations, including those with unknowns on both sides, is reviewed and practised throughout.

Models and representations

- Geoboards can be used to review angles in polygons prior to starting the chapter.
- Dynamic geometry software can be a quick way to recall and revise all the key facts and to see the angles changing as the transversal is moved.
- Bar models can help to model the equations and to structure students' solutions.
- Students should be encouraged to draw their own diagrams so that they can add in extra labels, etc. to support their reasoning.

Behind the questions

Are you ready?

These questions remind students about the angle facts that they will be using during the chapter, including those concerning angles in polygons. They also provide an opportunity to review simplifying algebraic expressions and solving equations.

Practice 9.3A

1 and **2** are both scaffolded questions. Students are asked explicitly about each fact that they need to use to solve the problems, and they are told to set up and solve an equation to answer each question. In question 1, they are given a choice of equations and asked to pick the correct one.

3 and **4** offer ample opportunities to put these skills into practice. Students need to set up their equations each time before solving them. The equations all involve unknowns on one side only.

5 introduces angles in parallel lines. The questions build up slowly so that algebra skills are developed before equations with unknowns on both sides are encountered later in the exercise. Students need to be able to identify the correct angle fact or they will not be able to set up the correct equation. They should also be reminded of the usefulness of substituting their answers back into the angle expressions to check that their answers are valid.

6 may look as if it will lead to simultaneous equations, given that there are two unknowns, but, in each example, students will be able to find the value of x before finding the value of y. This could be a good opportunity to address the fact that a single equation with two unknowns cannot be solved.

7 utilises the fact that exterior angles in a polygon add up to 360° as well as the relationship between interior and exterior angles.

8 extends to solving equations with unknowns on both sides. Bar models can be used to support students if needed. Again, the value of substituting the answer back into the expressions to check its validity should be emphasised.

9 provides a different context and reminds students about their work on constructions from Chapter 5. The equation that students need to solve will, again, have unknowns on both sides.

What do you think? 9.3A

1 needs a lot of thought. It's an excellent question to demonstrate that the fact being proved or shown cannot be used in the proof or demonstration! Here, students need to use the fact that the line ABC is straight to determine the value of x before checking to see whether angle DEB is co-interior to angle ABE, and hence to prove or disprove that the lines are parallel.

2 is a little harder than the questions in the practice exercise since the transfer between the given fact and algebraic notation is more challenging.

3 is similar to question 1, in that students are using angle facts in reverse to determine whether or not the lines are parallel. There are different ways to approach this and it would make a good discussion task if time allows.

Consolidate – do you need more?

This section provides further opportunities to review setting up and solving equations using a range of different angle facts. It could be a homework task or used for revision at a later date.

Stretch – can you deepen your learning?

1 needs a bit of algebraic manipulation. Bar models may be particularly helpful here. Conceptually, this is not challenging but is likely to be quite different to what students have seen before.

2 is a "show that" question. The emphasis should be on clarity of written work, which will set students up well for the next chapter.

3 is a potential investigation that can be worked on in pairs. First, students can explore the conjecture numerically before considering how to prove it algebraically.

Reflect

This encourages reflection on different approaches to solving equations; bar models and the use of other manipulatives can be explored.

9.4 Geometric conjectures

Small steps

- Conjectures with angles
- Conjectures with shapes
- Link constructions and geometric reasoning (H)

Rationale

In the final chapter of this block, students will develop their reasoning skills by exploring conjectures, proofs and demonstrations in the context of angles and shapes. The first part of the chapter is concerned with angle facts and properties of shapes. Students can start gently by exploring simple conjectures such as "All triangles have a line of symmetry"; questions should be asked about how this could be proved or disproved, leading to the concept of a counterexample. Eventually the chapter moves to demonstration-type questions ("show that") before more formal proof.

The second part of the chapter explores constructions in more depth. This is a Higher step but the earlier questions are certainly accessible and enable students to review their construction skills.

Things to look out for	Interleaving opportunities
• There is a lot of new language here that needs to be reiterated: demonstration, proof, conjecture, counterexample, etc. A class glossary or a set of examples may be helpful. • Some students will struggle with the concept of a counterexample. • Many students find the concept of formal proof challenging. For abstract, algebraic questions, students could be asked to work on equivalent examples with numbers so that they can use this as a guide to support them when they move on to algebraic notation.	• Work on angles from Books 1 and 2 is reviewed. • Students will need to use facts about symmetry, diagonals, side lengths, etc. of shapes. • Work on constructions is explored in the second part of the chapter. • Algebraic skills are reviewed in the context of more formal proofs.

Models and representations

- Bar models can be used to demonstrate equivalences, such as that if $a + b = c$ and $a + b = d$ then $c = d$.

Behind the questions

Are you ready?

These questions review the angle facts that students have met so far. Students need to feel confident with the types of problems that involve chains of reasoning, so these are included as well.

Practice 9.4A

1 involves some simple conjectures about triangles. Students can work in pairs to sketch triangles and explore Mario's ideas. The idea of a counterexample follows naturally from this work and could be discussed at this stage.

2 is similar to the first question but involves conjectures about angles. Again, this is a good opportunity to discuss ideas and how a conjecture can be proved or disproved.

3 is a scaffolded example that shows students how to set out their working by giving reasons for each step and finally drawing a conclusion. This can be used as a model for question 4, which is very similar but no step-by-step support is given.

5 formalises the idea of a counterexample.

6 could form an investigation in which students consider when Chloe's conjecture holds and when it does not.

7 provides an accessible set of statements that students can explore. Rather than simply being true or false, some of the statements are "true if …", which adds a layer of complexity to the work so far.

8 is a "show that" question similar to those met earlier in the chapter. It is numerical and uses angle facts about triangles and straight lines. Students should be reminded to give reasons for each stage of their working.

9 is a chance to review algebra skills. When finding the value of *x*, students need to find the angles in the parallelogram and then justify how this shows that the shape is a parallelogram. They may need some support with this last step.

10 is another "show that" question. These will be built on for those students who want to tackle more abstract proofs in the "What do you think?" section.

What do you think? 9.4A

1 uses angles in parallel lines as a vehicle to develop more abstract proof. Students should be encouraged to sketch the diagram and to use the hint given.

2 can seem difficult to get started on. Students should be encouraged to work out and add in any angles that they can. They should also think about which angle facts apply in this context. The fact that they need to prove is derived from the angles of triangle ABD adding up to 180°.

3 is more challenging since the relationship to be derived is not given. Students need to explore the problem by working out whichever angles they can (in terms of *x* and *y*) and then finding an angle fact that uses both.

Practice 9.4B

1 is a good opportunity to review constructing a perpendicular bisector before exploring the properties of the shape made when the quadrilateral ABCD is created.

2 explores drawing an angle bisector. There is also a chance to review congruent triangles.

3 reminds students about the connection between the construction of a 60° angle and an equilateral triangle.

4 can be a really good task to get students talking and trying to verbalise why all of the cases fail. It serves as a good reminder about the key properties of different types of triangles.

5 begins by exploring the perpendicular bisectors of the chords of a circle. The idea is then reversed to show that given three points (and hence two chords), a circle can be constructed.

What do you think? 9.4B

1 informally explores circle theorems, reviewing constructions, shape properties and congruence at the same time. There is no need to formalise the theorems; this should be seen as a chance to develop conjecturing.

2 At first glance it is not obvious why the areas are the same. Drawing the perpendicular from A to DB shows that both triangles have the same height, and knowledge of the diagonals of a parallelogram shows that the bases must also be equal.

Consolidate – do you need more?

This section provides further opportunities to consolidate and review the content from this chapter.

Stretch – can you deepen your learning?

1 is a much more challenging proof than those met so far. Again, students should start by writing down everything that they can work out and then look for connections.

2 could be used as an investigation and gives a chance to review angles in polygons in more depth.

3 looks at two ways of constructing a kite and gives students a chance to develop their reasoning skills by explaining why both methods work and how the diagrams can be finished off.

Reflect

1 reminds students of the language of proof and demonstration.

2 is a chance for students to formalise their understanding of a counterexample.

10.1 Symmetry

Small steps

- Identify the order of rotational symmetry of a shape
- Compare and contrast rotational symmetry with line symmetry

Rationale

In this chapter, students learn about rotational symmetry for the first time. This is a good topic to link to real-life examples such as the patterns within car wheels, road signs, flowers and snowflakes. Students need to understand the difference between regular and irregular shapes so that they can appreciate how this affects the order of rotational symmetry.

Things to look out for

- Some students may think that a shape can have order of rotational symmetry 0.
- Assuming that the order of rotational symmetry is always the same as the number of lines of symmetry.
- Assuming, for example, that all hexagons have order of rotational symmetry 6 because regular hexagons do.
- Some students may not put the pencil exactly in the middle of the shape when using tracing paper to determine the number of times it fits its outline when rotated; this will give incorrect answers.
- Not taking into account a pattern or shape within a shape, for example when looking at the order of rotational symmetry of road signs.

Interleaving opportunities

- This is a good opportunity to revisit names and properties of 2-D shapes, including regular and irregular shapes.
- Students should be able to describe and carry out clockwise, anticlockwise, quarter, half and three-quarter turns.
- The concept of turning can be taken further by looking at the angle of each turn, for example 90°, 180° and 270°, or even further by considering angles in regular pentagons, hexagons, etc.
- Higher students can make links to sums of interior and exterior angles.

Models and representations

- Tracing paper can be used to help students to identify the order of rotational symmetry. It is useful to draw an arrow at the starting point before rotating the paper about the centre of the shape.

Behind the questions

Are you ready?

These questions revisit some key skills to boost students' confidence before working through the chapter. If students cannot name all the 2-D shapes in question 3, you may need to revise these, for example in starter activities. Regular practice will help students to retain this knowledge. Students may think that the shapes in parts ii and iii are regular because they have hatch marks, and they may not be as familiar with isosceles trapeziums as they are with isosceles triangles. A good topic for discussion is whether there are any other shapes that can be called isosceles: triangles and trapeziums are the only ones.

Practice 10.1A

1 focuses on the order of rotational symmetry for regular shapes. Students should notice that the order of rotational symmetry is equal to the number of sides.

2 addresses the misconception that the rule discovered in question 1 applies to all shapes. For example, a rectangle has four sides but the order of rotational symmetry is 2 not 4. You could ask, "Do you think this is because a rectangle has two pairs of equal sides?"

3 looks at less familiar shapes.

4 gives students the opportunity to consolidate their understanding about the difference between the orders of rotational symmetry for regular and irregular shapes with the same number of sides.

5 invites students to work in reverse and draw a shape given its order of rotational symmetry. Students should compare their answers and approaches with a partner's.

6 looks at rotational symmetry in a real-life context. This could be extended to considering rotational symmetry of other road signs.

7 looks at rotational symmetry in another different real-life context. Students could take this further by designing their own car wheels with rotational symmetry of a given order.

What do you think? 10.1A

Question 1 introduces the rotational symmetry of a circle. Students should realise that even after being turned through a fraction of a degree the circle will look the same. Question 2 considers the effect of drawing one shape inside another. As extension work, students could consider variations of this, such as "What if the shapes do not share the same centre?" or "What if there is a pattern on one or both of the shapes?"

Practice 10.1B

1 asks students to compare the number of lines of symmetry of regular shapes with their order of rotational symmetry. They should recognise that, for regular shapes, the number of lines of symmetry is equal to the order of rotational symmetry, which is also equal to the number of sides.

2 considers irregular shapes. These carefully chosen examples will ensure that students don't assume that the number of lines of symmetry is always equal to the order of rotational symmetry.

3 encourages students to reflect on the similarities and differences between questions 1 and 2.

4 allows students to see that irregular shapes can have lines of symmetry. They should be encouraged to compare answers with a partner.

5 addresses the common misconception that irregular shapes always have rotational symmetry of order 1.

What do you think? 10.1B

Question 1 looks at symmetry in patterns drawn on squared grids. Students could design their own patterns and consider the similarities and differences between them. Question 2 addresses a common misconception that arises when students generalise too quickly.

Consolidate – do you need more?

These questions provide further opportunities for students to consolidate working with symmetry. If further practice is needed, students could work in pairs to create their own similar questions.

Stretch – can you deepen your learning?

1 interleaves learning on fractions and percentages. Students could be challenged to consider how the font or handwriting would affect the answers.

2 and **3** require students to use their knowledge of the angle sums of polygons to identify symmetry in two regular shapes.

Reflect

Students are encouraged to reflect on their learning by describing the similarities and differences between rotational symmetry and line symmetry in their own words.

10.2 Rotation

Small steps

- Rotate a shape about a point on the shape
- Rotate a shape about a point not on the shape

Rationale

In this chapter, students learn about rotation. As shown by the small steps, the learning has been split into two main learning points – rotation about a point on a shape and rotation about a point not on a shape. Rotation about a point on a shape is often skimmed, and students don't always have the opportunity to answer questions of this type. This chapter ensures that students have sufficient practice at answering both types of questions and appreciate what's the same and what's different when the centre of rotation is on a shape and when it is not. Students will be asked to perform and describe rotations both on and off a coordinate grid.

Things to look out for

- Getting clockwise and anticlockwise mixed up.
- Not understanding how far a 90°, 180° or 270° turn is.
- Students may use tracing paper to rotate the shape correctly but then may redraw it inaccurately.
- Students may plot the centre of rotation incorrectly.
- A common error is to forget to state all the key points when describing a rotation, for example omitting the direction of turning or the centre of rotation.

Interleaving opportunities

- Students will use their knowledge of plotting coordinates.
- Students can link quarter turns, half turns and three-quarter turns with the correct angles.
- The names of a variety 2-D shapes will be revised throughout the chapter.
- Students can practise accurate drawing of 2-D shapes and using coordinate grids with appropriate scales.
- Students will need to recognise when shapes are congruent.
- This chapter provides an opportunity to practise using correct mathematical language, e.g. "vertices", when describing shapes.

Models and representations

- You can use a clock to remind students about which direction is clockwise and which is anticlockwise.
- Provide students with tracing paper to help them to perform and describe rotations.

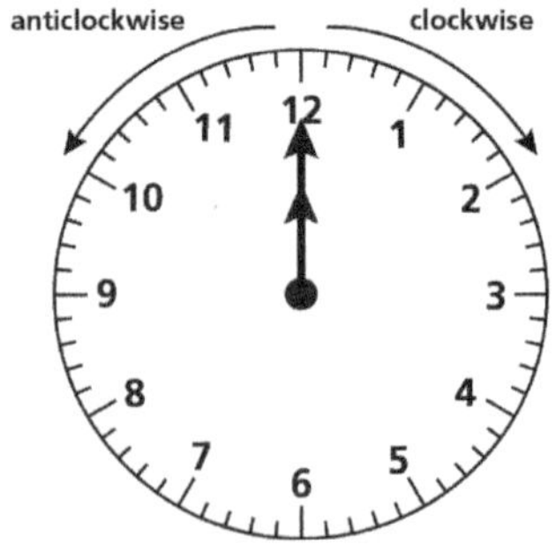

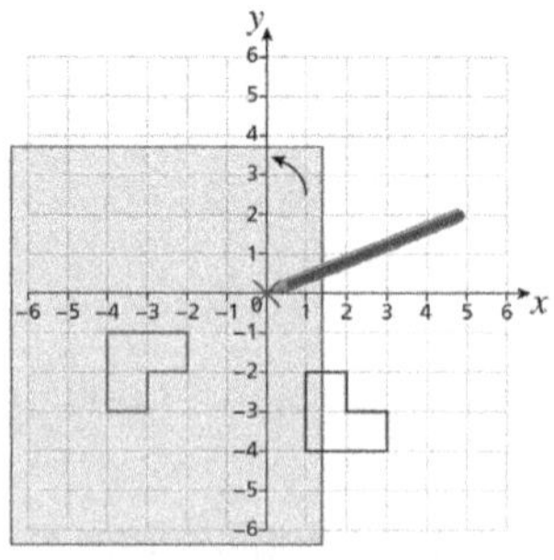

Behind the questions

Are you ready?

These three questions cover the essential prerequisite knowledge for this chapter. Question 1 ensures that students can read coordinates; be aware that some students may write them in the wrong order. Question 2 revisits the idea of congruence; students should be able to explain in their own words what this means. Question 3 links descriptions of turns to the number of degrees that each turn represents; students could be encouraged to draw pictures to support their reasoning.

Practice 10.2A

1 ensures that students can rotate a shape through a given angle without a centre of rotation. It is essential that they can draw a shape in the correct orientation after a rotation. They could be asked to discuss what they notice about their answers to parts iii and iv for each question part, and why this happens.

2 explores common misconceptions that arise when performing transformations. Students should be able to explain each error and carry out the correct rotation.

3 allows students to practise rotating a shape about a point on the shape. Notice that the same rectangle is used in each part so that students can focus their attention on how the position of the centre of rotation affects the position of the image.

4 involves rotating a shape on a coordinate grid. The rectangle and the required turn are the same in each part so that students can focus on how the position of the centre of rotation affects the position of the image. To provide an extra level of difficulty, students need to plot each centre of rotation themselves; however, all these centres are still points on the shape.

5 This question uses the same coordinate grids as in question 4, but here the centre of rotation is no longer a point on the shape. Students should be encouraged to consider how the position of the centre of rotation affects the position of the image. They should also compare answers to corresponding parts in questions 4 and 5 to identify what's the same and what's different.

6 Students will need to articulate their findings from questions 4 and 5.

7 requires students to describe rotations that have been performed in one quadrant. They should recognise that any rotation can be described both in a clockwise and in an anticlockwise direction by means of appropriate angles and understand the relationship between these angles. They should also realise that a rotation of 180° can be carried out in either direction so it is not necessary to specify this.

8 provides practice at carrying out rotations in all four quadrants.

9 requires students to describe rotations fully. Stem sentences can be provided for support, for example "Shape B is a rotation of shape A by ___ degrees __________ about the point _____." Students can use these to structure their answers and ensure that they are using the correct mathematical language and give all the necessary information.

What do you think? 10.2A

These four questions require students to think more deeply about rotation. They consider how the number of degrees through which a shape is turned affects the image. In question 4, students should be encouraged to provide examples to support their reasoning.

Consolidate – do you need more?

These questions provide extra practice on rotating shapes for students who need it. If further consolidation is needed, students could work in pairs to create their own similar examples.

Stretch – can you deepen your learning?

1 Students could be encouraged to draw a diagram to support their thinking.

2 looks at the idea of invariant lines in an informal way. Investigating the link between the coordinates of a point before and after a rotation will help students when they formally study invariant points and lines in the future.

3 builds on the ideas covered in question 2.

4 Students should be encouraged to use examples to support their reasoning. They could compare their approach and their findings with a partner.

Reflect

Students reflect on their learning and describe it in their own words. It is important that students can identify the minimum amount of information needed to perform a rotation; this will ensure that they are able to state all the key features when asked to describe a rotation fully.

10.3 Translations and beyond

Small steps

- Translate points and shapes by a given vector
- Compare rotation and reflection of shapes
- Find the result of a series of transformations (H)

Rationale

The first section of this chapter focuses on both performing and describing translations. Although students will have encountered translations at KS2, they have only used instructions such as left, right, up and down to describe them. Many students will find this transformation difficult when they are introduced to vector notation for the first time. Comparing vectors with coordinates can help students to understand what the numbers represent, i.e. that the top number of the vector corresponds to the *x*-value as it describes a movement left or right and the bottom number corresponds to the *y*-value as it describes a movement up or down. In the second part of the chapter, students compare rotations and reflections before moving on to finding the result of a series of transformations. Although this is a Higher step, encourage all students to try some of the questions. These are good way of consolidating their prior learning.

Things to look out for

- Getting the numbers in the vectors mixed up. Often students will count up or down by the top number and across by the bottom number in the vector.
- Some students may miscount the squares, counting their starting point as 1 instead of 0.
- Students may translate one vertex of a shape correctly but then draw the rest of the shape incorrectly.
- Students may get confused if the resulting shape overlaps the original shape.

Interleaving opportunities

- Students will use their knowledge of plotting coordinates.
- Students can practice naming and accurate drawing 2-D shapes using appropriate scales.
- Students will need to recognise when shapes are congruent.
- This chapter provides an opportunity to practise using correct mathematical language, e.g. "vertices", when describing shapes.
- Students need to be able to draw and recognise lines such as $y = 1$, $x = -4$ and $y = x$.

Models and representations

- Students should use square grids to support their understanding of vector notation. These could be unmarked grids or sets of coordinate axes. Encourage students to notice what's the same and what's different about coordinates and vectors, discussing the different ways they are represented.

Behind the questions

Are you ready?

This question covers the essential prerequisite knowledge for this chapter. Students are asked to describe each movement between points using words – this will be familiar to them from previous work. They will build on this when they start to work with vectors later in the chapter.

Practice 10.3A

1 ensures that students can correctly interpret translation vectors. Notice that the numbers used in each question part are the same: however, their positions and signs change so that students can see the effect this has on the movement.

2 is a scaffolded question that looks at translation on a squared grid. Students are given a description of the translation in words, and in part a they are asked to write this as a vector. In part b, they are introduced to a common misconception and asked to explain the mistake. This is designed to prevent students from making this mistake themselves.

3 requires students to describe translations using vector notation. Stem sentences could be provided for support so that students can focus on writing the translation vector. Students could be encouraged to describe each translation both in words and as a vector to reinforce their understanding of the link between the two.

4 Here students practise performing translations that are given in vector notation. They could be encouraged to write the meaning of each vector in words first and to check that every vertex has been translated correctly.

5 Watch out for students assuming that Zach is correct because shape A is a translation of shape B by the given vector. They should pay close attention to the order of A and B in the question to understand the mistake that has been made; this will help them when describing translations later in the exercise.

6 enables students to consolidate their understanding of translation in all four quadrants. They could also consider the coordinates of each vertex before and after a translation and describe what they notice.

7 follows on from question 5 by requiring students to describe translations. Make sure that they are describing the translation from the object to the image and not the wrong way round. Encourage them to look back at question 5 before checking their answers. Stem sentences can be used to support students.

8 invites students to spot the connection between a translation vector and what happens to the coordinates of each vertex of a shape. They should be able to explain the reason for this in their own words.

What do you think? 10.3A

These two questions provide opportunities for students to think more deeply about translations. They should realise that the resulting shape after a translation is always congruent to the original shape.

Practice 10.3B

1 Here scaffolding is provided to enable students to perform a series of transformations on the same shape. Suggest that they label each shape as they go along to avoid mistakes. Students should be able to explain the key differences between the transformations and how this is manifested in the resulting images.

2 encourages students to reflect on the transformations they have seen so far and consider common misconceptions. Highlighting these explicitly can help students to avoid making them themselves.

3 highlights the effect of changing the order when performing a series of transformations.

4 and **5** provide practice at performing a series of transformations.

6 gives students the chance to practise describing transformations. When describing the result of a series of two or three transformations, students should be encouraged to draw the interim shapes.

What do you think? 10.3B

These two questions encourage students to think more deeply about a series of transformations. They should use examples and counterexamples to support their reasoning in question 2.

Consolidate – do you need more?

These questions provide extra practice at drawing translations and combining transformations. If more questions are needed, students could work together in pairs to create their own similar examples.

Stretch – can you deepen your learning?

1 and **2** consider how translation and reflection affect the properties of a shape. Encourage students to make predictions and then test them. They should use correct mathematical language, e.g. "congruent".

3 interleaves previous learning on algebra. Students should be able to form and solve equations in terms of *a* and *b* and explain each step of their reasoning.

4 Students could be encouraged to draw a diagram to support their thinking.

Reflect

Students reflect on their learning by describing the key differences between the different transformations in their own words.

11.1 Working with right-angled triangles

Small steps

- Squares and square roots (R)
- Identify the hypotenuse of a right-angled triangle
- Determine whether a triangle is right-angled

Rationale

In this chapter, students are introduced to Pythagoras' theorem for the first time. Initially, the focus is on identifying the hypotenuse of a right-angled triangle. Students are given triangles in different orientations so that they need to identify the hypotenuse as being opposite the right angle rather than getting used to it being in a certain position. In the second part of the chapter, students use Pythagoras' theorem to determine whether or not a triangle is right-angled. Correct mathematical language should be encouraged throughout, e.g. "This triangle is right-angled because the sum of the squares of the two shorter sides is equal to the square of the hypotenuse."

Things to look out for

- Students sometimes confuse finding the square root with halving. If this happens, prompt them to square their answers to demonstrate that this is incorrect.
- Students might get used to the hypotenuse being in a certain position if a right-angled triangle is always presented in a similar "standard" way. To avoid this, ensure that your examples show right-angled triangles in different orientations so that students recognise the hypotenuse as being opposite the right angle.

Interleaving opportunities

- Properties of shapes can be revisited.
- This is a good opportunity to recap converting between metric units by giving side lengths measured in different units.
- Geometric notation can be revisited by asking students to work out the length of AB, say, rather than finding the value of a single variable, e.g. y.
- Formulae for finding areas of shapes can be revisited. Although this chapter specifically looks at squares, other 2-D shapes can also be drawn on the sides of the triangles.

Models and representations

- A visual representation of Pythagoras' theorem can support understanding. This representation shows that when you draw a square on each side of a right-angled triangle, the area of the largest square is equal to the sum of the areas of the two smaller squares.

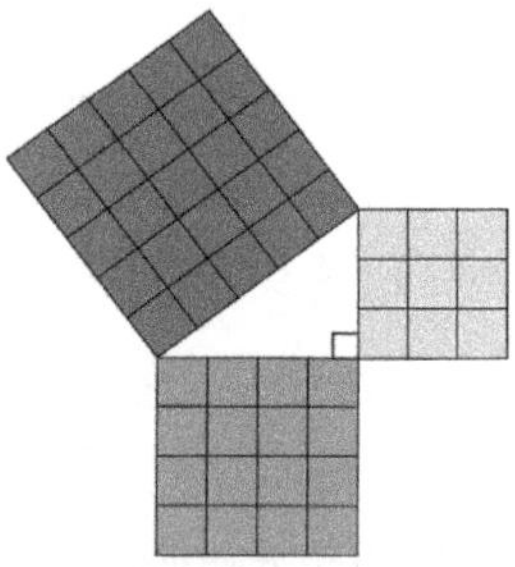

Behind the questions

Are you ready?

These four questions ensure that students have the essential prerequisite knowledge for this chapter. Students should be familiar with square numbers and square roots and should be able to calculate them confidently.

Practice 11.1A

1 helps students to consolidate their ability to construct triangles. They should be introduced to the term "hypotenuse" and know that this is the longest side of a right-angled triangle. Students should recognise that the hypotenuse is the side opposite the right angle.

2 builds on question 1 by asking students to identify the hypotenuse of some right-angled triangles drawn in a variety of orientations.

3 summarises what students have seen in the previous two questions and ensures that they are able to articulate what is meant by the hypotenuse of a right-angled triangle.

4 Here students reverse the process. Now they know that the hypotenuse of a right-angled triangle is opposite the right angle, they can use this to identify which angle of a triangle is the right angle.

What do you think? 11.1A

These three questions provide opportunities for students to consider the hypotenuse of a right-angled triangle in more depth.

Practice 11.1B

1 and **2** Here students start to apply Pythagoras' theorem and recognise its uses. They should be able to substitute values and articulate that when the sum of the squares of the shorter sides is equal to the square of the hypotenuse, then the triangle is right-angled. When it is not, then the triangle does not have a right angle.

3 points out that sometimes sketches are not drawn accurately but this does not mean that the triangle cannot be right-angled.

4 provides further practice at using Pythagoras' theorem to decide whether or not a triangle has a right angle. Students should be encouraged to explain their reasoning logically.

5 Students should notice that the units of the three side lengths are not the same, so some will need to be converted. Students could be asked to show that the triangle is still right-angled when working in mm, m or cm.

What do you think? 11.1B

These two questions encourage students to think more deeply about using Pythagoras' theorem. Question 1 highlights that this provides a more accurate way of finding a side length of a triangle than measuring. Question 2 links Pythagoras' theorem to similar triangles; students should be prompted to recognise that if one Pythagorean triple is known, then others can be found by considering similar triangles.

Consolidate – do you need more?

These four questions allow students to consolidate their understanding. They revisit identifying the hypotenuse or the right angle of a right-angled triangle and provide further practice at using Pythagoras' theorem to decide whether or not a triangle is right-angled. If more is needed, students could work in pairs to create similar examples of their own.

Stretch – can you deepen your learning?

1 This question introduces students to the idea of using Pythagoras' theorem on a coordinate grid. This will be built upon in more detail later in the block.

2 Students could investigate drawing different shapes on the sides of the triangle to recognise that they can all be used to demonstrate whether a triangle has a right angle. Why we tend to focus on squares and the reason for this could make an interesting discussion point.

3 builds on question 2 in the "What do you think?" section, in which students used similar triangles to find Pythagorean triples, but in a more formal way. As an extension, students could use algebraic notation to support their proof.

Reflect

Students are encouraged to reflect on their learning and summarise it in their own words. They could be encouraged to draw diagrams to support their reasoning.

11.2 Finding unknown sides

Small steps

- Calculate the hypotenuse of a right-angled triangle
- Calculate missing sides in right-angled triangles
- Explore proofs of Pythagoras' theorem

Rationale

In this chapter, students use Pythagoras' theorem in a more formal way. They were introduced to Pythagoras' theorem in the previous chapter and saw how it can be used to determine whether or not a triangle is right-angled. Here they build on this knowledge by calculating the lengths of unknown sides in right-angled triangles. This is the first time that they will have seen the formula for Pythagoras' theorem. Practice 11.2A focuses on calculating the hypotenuse and Practice 11.2B includes examples of calculating the length of a shorter side.

Things to look out for

- A common mistake is for students to always add the squares of the given values, even when calculating a shorter side. To avoid this, they could be encouraged to think about what they know about the unknown length before calculating it. For example, if it's the hypotenuse, then it will be longer than the other two sides; if it's not, then it will be shorter than the hypotenuse. This will help students to sense-check their answers and recognise when they have made a mistake.

Interleaving opportunities

- Knowledge of fact families and inverse operations is used when calculating the length of a shorter side of a triangle.
- Area and perimeter can be revised in questions where students first need to use Pythagoras' theorem to find a length.
- Students could apply Pythagoras' theorem to questions where the lengths are given as fractions and decimals.
- Students will use their knowledge of rounding throughout.

Models and representations

- A visual representation of Pythagoras' theorem can help students to understand why they add when finding the hypotenuse and subtract when finding a shorter side.
- As in many chapters in this book, bar models are useful, particularly in helping to rearrange formulae.

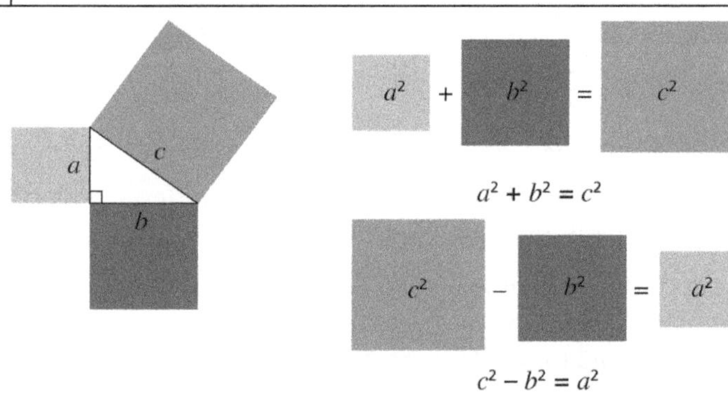

Behind the questions

Are you ready?

These four questions are designed to ensure that students have the knowledge needed for the rest of the chapter. Questions 1 and 2 involve finding the area of a square given its side length and vice versa. Questions 3 and 4 revisit learning from the previous chapter to ensure that students can correctly identify the hypotenuse of a right-angled triangle and use Pythagoras' theorem to show that a triangle is right-angled.

Practice 11.2A

1 Students use their knowledge of area to support their understanding of Pythagoras' theorem.

2 Here, students should be encouraged to explain the mistake based on both the algebraic method and in relation to question 1. They should be able to use knowledge of inverse operations to explain that 169 is x^2 not x and that the square root of this number needs to be found. They should also be able to explain that 169 represents the *area* of the largest square and not the side length of it.

3 Students could sketch squares on each side of the triangles to support their working but should be able to articulate why it would be inefficient to accurately construct each of these squares. You could have an interesting discussion around the given values not being Pythagorean triples.

4 Students may not recognise that the units given on the two side lengths are different and so one of them will need to be converted before the length of the hypotenuse can be found. Students could be encouraged to calculate the length of the hypotenuse in both mm and cm to show that both approaches will work.

7 provides opportunities to recap multiplication and division of fractions in the context of Pythagoras' theorem. Students may struggle to find the square root of a fraction and might need prompting, by means of some carefully chosen examples, to recognise that $\frac{a}{b} \times \frac{a}{b} = \frac{a^2}{b^2}$ so the square root of $\frac{a^2}{b^2}$ is $\frac{a}{b}$.

What do you think? 11.2A

These three questions provide opportunities for students to think more deeply about Pythagoras' theorem by applying it to slightly more challenging contexts. In part c of question 1, students might not recognise that they can form a right-angled triangle within the trapezium and could be encouraged to sketch it to allow them to visualise this. In question 2, students extend this idea to form right-angled triangles within a trapezium.

Practice 11.2B

1 Students should recognise that Bobbie's answer cannot be correct because the hypotenuse of the triangle is 13 cm, so the other two sides must be less than this.

2 Students should consider what they know about the unknown length before performing any calculation. For example, in part a they should realise that it must be less than 15.

4 A common mistake with this sort of question is that students multiply 6 by 8 then divide the answer by 2. This question is scaffolded in a way that tries to avoid this, but students could be encouraged to consider what mistakes they think people might make when answering this sort of question. Other common errors are labelling the base of the right-angled triangle as 6 or seeing the numbers 6 and 8 and assuming that the third value must be 10.

What do you think? 11.2B

Question 1 formally highlights the difference between the questions in Practice 11.2A and Practice 11.2B. In question 2, students should think carefully about what each of the lengths given in the question represents in their diagram. Question 3 provides an opportunity to look at both Perigal's dissection and an algebraic proof of Pythagoras' theorem. You may need to remind students how to expand a pair of binomials.

Consolidate – do you need more?

These two questions provide further opportunities for students to practise calculating the length of unknown sides using Pythagoras' theorem. They should be encouraged to get into the habit of always identifying the hypotenuse of the right-angled triangle first to help them to decide on the steps they need to take.

Stretch – can you deepen your learning?

1 and **4** Students could be encouraged to split their shapes and draw the right-angled triangles separately.

2 Students might need so be reminded of compass directions and the meaning of due north and due east.

3 Students could draw "before" and "after" diagrams to support their reasoning and organise their working.

5 Students could use algebraic notation to support their working or could, instead, go back to the idea of squares sketched on each side of the triangle to help them to decide what steps are needed.

6 Students may need a hint that the shape is a trapezium, and to compare the total area with the sum of the areas of the triangles.

Reflect

Students are encouraged to reflect on their learning from this chapter and explain it in their own words. They should be able to articulate that when using the formula $a^2 + b^2 = c^2$, the hypotenuse should always be labelled c but a and b can be either of the shorter sides.

11.3 Beyond triangles

Small steps

- Use Pythagoras' theorem on coordinate axes
- Use Pythagoras' theorem in 3-D shapes (H)

Rationale

In this chapter, students extend their knowledge of Pythagoras' theorem and apply it in a wider range of contexts. Visual representations will be essential for ensuring that students can see where right-angled triangles are formed within a more complex shape. Practice 11.3A focuses on using Pythagoras' theorem on a coordinate grid whilst Practice 11.3B focuses on working in three dimensions.

Things to look out for

- Students may find incorrect values for the horizontal and vertical distances between a pair of coordinates, particularly when working in four quadrants. Drawing clear diagrams and labelling them with the coordinates can help to overcome this problem.
- Students may use incorrect values when working in 3-D, when it is harder to spot right angles from a 2-D representation of a shape. Encourage them to draw each right-angled triangle they are working with separately from the main 3-D drawing and to label all vertices and side lengths so that links can be made back to the original diagram.

Interleaving opportunities

- Students can revisit their learning on the Cartesian plane. They are given opportunities to plot and read coordinates throughout Practice 11.3A.
- Translation vectors can be revisited, and Pythagoras' theorem can be used to find the distance that a shape has moved.
- Practice 11.3B provides opportunities to revise properties of 3-D shapes.
- Algebraic notation can be interleaved throughout, in particular when students are generalising their findings. This is explored in the "Stretch" exercise.

Models and representations

- Geoboards can be used to represent points and line segments to support working on a coordinate grid.

Behind the questions

Are you ready?

These three questions cover the essential prerequisite knowledge for this chapter. Questions 1 and 2 recap learning from earlier in the block: they ensure that students can calculate missing lengths in right-angled triangles using Pythagoras' theorem. Question 3 checks that students are able to find the lengths of horizontal and vertical line segments on a coordinate grid. Students could be encouraged to explain how to find the length of a line segment both by using the grid and by using only the coordinates.

Practice 11.3A

1 builds on question 3 from the "Are you ready?" exercise. In parts a and b, students should be able to use both the coordinate grid and the coordinates of each point to explain their answers. In part c, students might find it useful to sketch the diagram and draw on the right-angled triangle to visualise the problem. This is a "show that" question, so students know the answer they are working towards and need to focus on writing a full method. This should enable any misconceptions to be identified early on.

3 In part a, students could be encouraged to make copies of the grid for each part for support. In part b, students might need to be reminded that the origin has coordinates (0, 0) and could be prompted to recognise why this means that the shorter lengths of the right-angled triangle are 11 and 10. In part c, students may wish to plot point F on a copy of the coordinate grid. Alternatively, they could explain how the question can be answered by just using the coordinates of points B and F without plotting them.

4 Building on question 3, students now need to calculate the length of a line segment with no visual support. However, students could draw a coordinate grid if they find this difficult. Some students may need to draw the grid accurately, whilst others may just need a sketch to help them to organise their thinking.

5 and **6** follow a similar structure to questions 3 and 4 but use all four quadrants. Working with directed numbers can often cause difficulties, and students should pay close attention to the minus signs to make sure that they don't make any numerical errors.

7 Students need to decide which two pairs of vertices would be opposite each other and should be able to deduce this using the *x*- and *y*-values of the given pairs of coordinates. They could be challenged to show that it doesn't matter which pair of opposite vertices they choose to calculate with – the length of the diagonal would be the same.

What do you think? 11.3A

Question 1 links back to previous learning on translations. Students should recognise that the translation vector determines the distance between X and X′ and so the coordinates of point X are irrelevant. Question 2 revisits previous learning on ratio. Students could be encouraged to first consider whether AC will be longer or shorter than AB and be able to reason why.

Practice 11.3B

1 This question is designed to get students into good habits when working with Pythagoras' theorem in 3-D. It can often be a conceptually challenging topic but, through the use of clearly labelled diagrams and a structured approach, students should realise that there isn't much difference from working in 2-D: often just an extra step is required.

2 Students should sketch the relevant triangles and label them clearly.

3 Students need to decide for themselves what steps to take in and what order. Diagrams are, again, useful.

4 Notice how FD isn't actually shown on this diagram. Students should be prompted to add this to a sketch of the cube for visual support.

What do you think? 11.3B

These three questions look at some different 3-D shapes. Students need to consider the information they are given, what they want to find out and the steps they need to take.

Consolidate – do you need more?

Questions 1 and 2 focus on using Pythagoras' theorem on a coordinate grid whilst questions 3 and 4 look at three-dimensional problems.

Stretch – can you deepen your learning?

1 and **2** These use the skills from Practice 11.3A but in a slightly different way. Students might not immediately see that several answers are possible for each question, but diagrams should assist with this.

3 prompts students to generalise their learning from Practice 11.3A. As an extra step in between parts b and c, if needed, they could be asked to find the distance between, say, (2, 7) and (*x*, *y*).

4 Students should avoid rounding too early to ensure that their final answers are as accurate as possible.

5 Students could test that their expression is correct by substituting values from earlier questions.

Reflect

Students reflect on their learning and describe it in their own words. In question 1, students should recognise that coordinates can always be used to find horizontal and vertical distances between points so Pythagoras' theorem can always be used to find the direct distance between the points. In question 2, students should compare their learning from Practice 11.3B to that from previous chapters in the block and reflect on the similarities and differences.

12.1 Calculating with enlargement

Small steps

- Recognise enlargement and similarity
- Work out missing sides and angles in a pair of given similar shapes

Rationale

In this chapter, students are introduced to the concepts of enlargement, similarity and scale factors. There is no need to formally teach enlargement on a grid at this stage: this will be covered in the next chapter. Students need to be confident in multiplying fractions by integers and working with ratio. These key concepts can be revisited from Book 2 if necessary.

Things to look out for	Interleaving opportunities
• Some students try to use additive relationships rather than multiplicative ones. • Students often want to divide when a shape becomes smaller rather than multiplying by a fraction or decimal. • Students may assume that angles increase or decrease in size as well as side lengths when a shape is enlarged.	• Multiplying fractions by integers • Working with ratio • Working with decimals • Multiplication by integers, decimals and fractions • Area and perimeter • Accurate drawing using rulers, compasses and protractors

Models and representations

- It is important for students to see enlargement and similarity in real-life contexts.

Behind the questions

Are you ready?

These four questions cover the essential prerequisite knowledge for the chapter to ensure that students can work confidently with fractions and ratios. Questions 1 and 2 are linked: prompt students to recognise the similarities and differences between calculating a unit fraction of 30 and a non-unit fraction of 30. In question 2, students should be able to explain why the answer to part a is less than 30 but for part b it is greater than 30. Questions 3 and 4 focus on writing ratios and fractions in their simplest form. Notice that the numbers used in corresponding parts are the same so that students can link the two formats.

Practice 12.1A

1 Part a checks that students can explain what it means for two rectangles to be similar, whilst part b highlights that similar shapes might not be presented in the same orientation.

2 Students should notice that only the base of the triangle changes, not the height, so these two triangles are not similar. It is common for students to assume that all isosceles triangles are similar, and this question provides a good opportunity to address this misconception early.

3 Students should realise that there are several answers for each of these parts; encourage them to compare their answers with a partner. You could suggest that they label the dimensions using different units or include decimals to challenge them further.

4 In part c, watch out for students saying that the scale factor of the enlargement is 2 because they go from the smaller shape to the larger one. If this happens, prompt students to compare parts a and c so that they appreciate why this cannot be the case.6

5 Students need to find the scale factor for an enlargement where the shapes are not presented on a squared grid. Encourage them to show each step of their working to establish a method that they can follow when working with fractional scale factors.

6 draws students' attention to the link between scale factors and ratios.

7 In part c, some students may multiply the angle by the scale factor 2. If this happens, ask students to find the sum of the angles in their enlarged triangles so that they realise why this can't be correct. The scale factor of enlargement here is not an integer; students could use ratios, as in question 6, to help them find the missing length.

8 interleaves learning on Pythagoras' theorem from Block 11. You could prompt students to realise that the 1.25 cm and 12 cm are not corresponding side lengths by using the scale factor found using the hypotenuse of each triangle.

9 Students should be able to state which sides in each triangle correspond and justify this.

What do you think? 12.1A

These three questions provide opportunities for students to think in depth about calculations involving enlargement. Question 1 involves multiplying by powers of 10 and could be extended by asking students to give their answers in metres or kilometres. In question 2, students might need to be reminded about the way in which angle notation is used to show which angles correspond in each triangle. Students could take this further by finding all the missing side lengths and then the perimeter of each triangle. They could also be asked to construct them accurately and measure the angles. Question 3 guides students towards realising that there will be several possible answers if they cannot which are the corresponding sides.

Consolidate – do you need more?

These questions provide further opportunities to perform calculations involving enlargement for students who require extra practice. They could also be encouraged to work in pairs to create similar examples of their own.

Stretch – can you deepen your learning?

1 Students should be able to explain why ABC and ADE are similar as well as calculate the length of AD. Encourage them to draw the two triangles out separately and label all the information they know on their diagrams.

2 Watch out for students attempting to use two of the given lengths to calculate the scale factor of enlargement. Instead, they should realise that they first need to find the hypotenuse of the smaller triangle using Pythagoras' theorem.

3 In part a, students could draw triangles ABC and CDE separately and label all the dimensions they know to support their understanding. Part b explores several common misconceptions surrounding similar triangles, and students should be stretched to consider how these mistakes could be avoided.

4 interleaves algebraic notation in the context of similar shapes.

Reflect

These two questions ensure that students understand the key learning points of this chapter. Students need to explain in their own words what is meant by similarity, and they should be able to give examples of scale factors that make a shape bigger as well as those that would result in a shape becoming smaller.

12.2 Enlarging a shape (1)

Small steps

- Enlarge a shape by a positive integer scale factor
- Enlarge a shape by a positive integer scale factor from a point

Rationale

This chapter focuses on positive integer scale factors only. The learning is split into two sections so that enlargements that can be performed anywhere and enlargements that are performed from a specific point are covered separately. Students first explore shapes for which all sides are vertical or horizontal and then they move on to shapes that have diagonal sides. Students should use the grid squares to help them to enlarge diagonal lines successfully: for example, if the movement is 3 squares to the left and 2 squares up when going from one end of a line to the other, then when the line is enlarged by scale factor 2 this movement will become 6 squares to the left and 4 squares up. Students often find enlarging a shape from a given point more difficult than simply enlarging a shape anywhere, so ensure that they are fully confident in enlarging the shapes given in the first section of the chapter before moving on to the second section.

Things to look out for

- Enlarging only one side of a shape.
- Inaccurate drawing.
- Some students get confused with diagonal sides and guess where these should be drawn. Encourage them to draw the enlarged horizontal and vertical sides first.
- Ignoring the centre of enlargement and just drawing the enlarged shape anywhere.
- Drawing the enlarged shape so that it touches the centre of enlargement.

Interleaving opportunities

- Working with coordinates in all four quadrants
- Knowledge of the properties of a variety of 2-D shapes
- Working with multiplicative relationships
- Accurate drawing
- Using correct mathematical language associated with 2-D shapes
- Working with negative numbers

Models and representations

- Students should be encouraged to draw “ray diagrams”.

Behind the questions

Are you ready?

In question 1, students might be unsure of the orientation in which they need to draw the rectangle. They should realise that this doesn't matter as long as the sides measure 4 squares and 3 squares. Questions 2 and 3 revisit the learning from the previous chapter. Students should recognise that the resulting shape in question 2 is similar to the original one. Question 3 is a reminder of what is meant by the term “scale factor”.

Practice 12.2A

1 addresses a common misconception whereby only one side of a shape is multiplied by the scale factor.

2 allows students to practise enlarging a shape by a given integer scale factor. Notice that all the shapes given are rectilinear; students will work with shapes having diagonal sides later in the exercise. Remind students to check that they have enlarged all of the side lengths by the same scale factor.

3 introduces shapes with diagonal sides, which adds another level of difficulty to drawing an enlargement. Students could compare methods with a partner and consider different ways in which the enlargement could be completed.

4 Students will probably find the left-hand shape easier because two of the sides are perpendicular. Enlarging the right-hand shape is trickier and there are several possible approaches.

5 asks students to enlarge more complex shapes.

6 requires students to reverse the process and find the scale factor of the enlargement. This should be linked to the learning from the previous chapter.

What do you think? 12.2A

These two questions provide opportunities for students to think more deeply about enlargement. They should realise that an enlargement by a scale factor of 1 would not change the shape. In question 2, students should test this conjecture using a range of different shapes. You could suggest that they label the side lengths algebraically and use algebra to show that the conjecture will always be true.

Practice 12.2B

1 Now students are introduced to enlargement from a specific point. This question addresses the misconception that one vertex of the enlarged shape will be on the centre of enlargement. Students should be able to explain that the distances of all vertices of a shape from the centre should also all be enlarged by the scale factor.

2 provides practice at enlarging shapes from a point. Here the scale factor is consistent for each part of the question so that students can focus on the different distances between the shape and the centre. Students may realise that they can complete each shape once the correct position of just one vertex has been found; they could be asked to explain how they know that each of the other vertices is in the correct position.

3 uses examples where the movement from the centre to some vertices of the shape is only horizontal or only vertical. Students should realise that if the horizontal/vertical movement from the centre to a vertex on the object is 0, then the horizontal/vertical movement to the position of the corresponding vertex on the enlargement will also be 0.

4 provides plenty of practice at enlarging a shape on a squared grid. Printed coordinate grids could be provided if necessary. Encourage students to clearly mark their centre of enlargement and show all their working.

5 Stem sentences could be provided to help students to structure their descriptions. For example, they could be asked to copy and complete “Shape B is an enlargement of shape A by scale factor ____ about the point ___”. This will ensure the use of correct mathematical language and that all the key information is given. Students may find locating the centre of enlargement difficult; encourage them to compare methods with a partner.

What do you think? 12.2B

In question 1, the centre of enlargement is inside the shape. In question 2, students are given the centre of enlargement and the image and they need to locate the original shape. You could also ask them to state the scale factor.

Consolidate – do you need more?

These questions allow students to practise enlarging shapes both with and without a centre of enlargement.

Stretch – can you deepen your learning?

1 links previous learning on translations with enlargement from a point. The question guides students towards realising that when a shape is enlarged from a centre, then the vector that describes the movement from the centre to each vertex is also multiplied by the scale factor.

2 Students could draw this or use the idea of translation vectors from the previous question. They could be encouraged to compare different methods and discuss the pros and cons of each.

3 looks at the misconception that the area of a shape is multiplied by the scale factor when it is enlarged. Although it does not need to be formally covered here, students could explore this further and discover that the area scale factor is the length scale factor squared.

4 You could initially assign values to p and q to help students to get started. Then suggest that they write down the translation vector from the centre to each vertex and use this to complete the enlargement.

Reflect

Students reflect on their learning from this chapter and explain it in their own words. They should describe the similarities and differences between enlarging with a centre and enlarging without a centre.

12.3 Enlarging a shape (2)

Small steps

- Enlarge a shape by a positive fractional scale factor
- Enlarge a shape by a negative scale factor (H)

Rationale

Here, students build on their learning from the previous chapter by enlarging shapes using fractional and negative scale factors. From their work on similarity, students will be aware of the fact that a fractional scale factor less than 1 results in a shape becoming smaller, and they should be able to transfer this knowledge to enlargement with a fractional scale factor from a point. The second part of the chapter focuses on negative enlargements; previous learning on translation and vectors will be useful here.

Things to look out for	Interleaving opportunities
• A common mistake is to draw the enlarged shape so that one vertex is at the centre of enlargement. • When enlarging with a negative scale factor, students often work out the position of one vertex and then draw the shape in its original orientation. Stress the importance of finding the new position of all vertices of a shape before drawing it to fully appreciate how a negative scale factor affects the orientation.	• Knowledge of similar shapes will be used throughout. • Multiplication by fractions is needed when working with fractional scale factors. • Translation vectors can be used to find the position of the enlarged shape when enlarging relative to a centre. • Revise area and perimeter.

Models and representations

- Ray diagrams are useful when describing enlargements.

Behind the questions

Are you ready?

These three questions check that students can confidently work with fractions. In question 1, students may need to be reminded that multiplying a fraction by an integer is the same as finding that fraction of the integer; you could ask them to compare these approaches. Question 2 reinforces the link between finding a fraction and division. Students may choose to work out the left-hand side first and then calculate the missing value on the right-hand side. However, they should notice that the missing value in each case is the reciprocal. Students could use knowledge of inverse operations to find the missing values in question 3.

Practice 12.3A

1 allows students to practise enlarging a shape by a fractional scale factor. No centre of enlargement is given in this question so students can just focus on how the side lengths of the shapes change. Notice that each of the scale factors is a unit fraction.

2 The shapes and scale factors are the same as those given in question 1, but here students are also given a centre of enlargement. As they already know what each enlarged shape should look like, students can focus on drawing it in the correct position.

3 Here the scale factor is a non-unit fraction. In part a, students should recognise that the size of the shape is correct as each of the side lengths is $\frac{2}{3}$ of the length of the corresponding side on the original shape. Part b exposes the common misconception of drawing one of the vertices of the enlargement at the centre. In part c, students can focus on finding the correct position of each vertex on the enlargement, as they already know from part a how big it should be.

4 provides further opportunities to practise enlarging a shape by a fractional scale factor on a coordinate grid.

5 requires students to describe enlargements. The use of stem sentences could help students to structure their explanations. Watch out for students describing the wrong enlargement as it common to want to describe how to map the smaller shape onto the larger one.

What do you think? 12.3A

These two questions provide opportunities for students to think more deeply about enlargement with fractional scale factors. In question 1, students need to use their knowledge of fractions greater than 1 to explain why the resulting shape will be bigger than the original. Question 2 interleaves previous learning on ratio, and students should make connections between the scale factor and each of the ratios.

Practice 12.3B

1 Here students use their existing knowledge and a process of elimination to decide which of the enlargements has a negative scale factor. Encourage students to explain in their own words what makes an enlargement by a negative scale factor different from an enlargement by a positive scale factor.

2 Earlier in the block, students saw that a scale factor of 1 leaves the shape unchanged. This, combined with their learning so far in this exercise, should lead them to the conclusion that the scale factor here is −1.

3 provides opportunities for students to perform enlargements by negative scale factors. Encourage them to consider the movement from the centre of enlargement to every vertex, rather than finding the new position of just one vertex and then making assumptions about the orientation of the enlargement.

4 and **5** provide further practice at performing enlargements by a negative scale factor. Here the shapes are drawn on a coordinate grid. Question 4 focuses on one quadrant only whilst question 5 works in all four.

6 Students should recognise that the enlarged shape is twice the size but on the opposite side of the centre of enlargement to support their reasoning of why the scale factor is −2.

What do you think? 12.3B

Question 1 looks at the relationship between rotation and enlargement by a negative scale factor. Students could be challenged to consider whether an enlargement by a scale factor such as −2 could be described in any other way and why this might be possible. Question 2 links enlargement with translation and introduces students to how vectors can be used to perform enlargements. Question 3 builds on these ideas.

Consolidate – do you need more?

These questions provide further opportunities to practise performing enlargements by fractional and negative scale factors both with and without a centre of enlargement. In addition, students could work in pairs to create similar examples of their own.

Stretch – can you deepen your learning?

1 You could suggest that students draw these enlargements to enable them to visualise the problem.

2 Encourage students to give examples and counterexamples to support their answers.

3 Students should recognise that there are two reasons why this is incorrect: a length can only be positive and the product of two negative numbers is positive.

4 builds on the idea of using translation vectors as a method of performing enlargements that was covered in questions 2 and 3 in the second "What do you think?" exercise. This could be extended to considering what the vectors would be if the scale factor was, for example, *p*.

5 You could suggest that students first assign values to *a* and *b* to help them to spot patterns before generalising.

6 introduces the idea of an invariant point. Students could be challenged to consider other transformations that might have a similar result.

Reflect

Students are asked to reflect on their learning from this chapter and explain it in their own words. They should be encouraged to give examples and counterexamples to support their reasoning.

12.4 More similarity (H)

Small steps

- Solve problems with similar triangles (H)
- Explore ratios in right-angled triangles (H)

Rationale

Building on their learning from the rest of this block, students now explore similar triangles in more detail. They will appreciate that triangles with equal angles are similar and solve problems that involve finding missing sides and angles. As this is Higher step, there is some challenging material such as GCSE-style questions with embedded triangles. Ratios both between and within triangles are considered, leading to an introduction to the trigonometric ratios. This chapter focuses on understanding trigonometry as an extension of similar triangles to right-angled triangles. The trigonometric ratios are used in 30-60-90 and 45-45-90 triangles only. This topic will be developed further in Chapter 17.3, or you may prefer to start Year 10 with an in-depth study of trigonometry, as suggested by the White Rose Maths KS4 schemes of learning.

Things to look out for

- Thinking that angles are increased/decreased by the same scale factors as the sides of a shape. Students should be able to prove that this is not the case by using their knowledge of the angle sum of a triangle.
- Students may not realise that two equal angles is enough to show the similarity of a pair of triangles. Again, students should be reminded of the angle sum property.
- Students may incorrectly identify corresponding sides. Encourage them to look carefully at the given angles and use the notation recommended in the Student Book for matching the letters of corresponding vertices. (If the triangles don't have letters, then students could be invited to add their own labels.)
- Students may be overwhelmed by the amount of new vocabulary in the trigonometry section. It is not expected that students should remember these, so it is useful to provide word cards and displays for support.

Interleaving opportunities

- This builds on the learning from Chapters 12.1, 12.2 and 12.3.
- Forming and solving equations is used throughout this chapter, in particular, those that start in fractional form and require rearrangement.
- Some questions require Pythagoras' theorem to find the lengths of unknown sides, and the term "hypotenuse" is crucial to the trigonometry section.
- Constructing triangles can be revisited.
- More general trigonometry is covered in Chapter 17.3.

Models and representations

- Students can use dynamic geometry software to produce and enlarge triangles to demonstrate that the ratios of the sides remain constant as the size changes.
- Students can construct and cut out sets of similar triangles to demonstrate that each angle of a smaller shape "fits" the corresponding angle of a larger shape.

Behind the questions

Are you ready?

Question 1 reminds students of the effect of an enlargement on side lengths and question 2 checks that students no longer have the misconception that the angles change when a shape is enlarged. Question 3 is fluency practice in basic similarity. Any students who have difficulty answering these questions may need intervention before moving on to the Higher steps covered in this chapter.

Practice 12.4A

1 and **2** check that students can identify corresponding sides and angles in pairs of triangles that are known to be similar.

3 develops this to include finding the lengths of unknown sides. Students also become aware that, because of the angle sum property of a triangle, two equal angles is enough to show that two triangles are similar.

4 Students will need to show that all three pairs of corresponding sides are in the same ratio to prove that the triangles are similar. They need to be confident with both types of proof – using ratios and showing that two triangles have equal angles.

5 and **6** provide increasingly complex fluency practice at using the similarly of triangles.

7 and **8** require knowledge of angle facts related to parallel lines to prove that the triangles are similar before going on to find the lengths of the unknown sides and the perimeter.

What do you think? 12.4A

1 challenges students to think about the difference between similarity and congruence.

2 requires the forming and solving of equations involving fractions to find the missing sides.

3 looks explicitly at the ratios of side lengths within a triangle, which will lead on to the trigonometric ratios in the next section.

4 builds on question 3 and makes use of the angle sum of a triangle within right-angled triangles.

Practice 12.4B

1 to **3** introduce the concept of the cosine of an angle by using the ratio of the side adjacent to the 60° angle and the hypotenuse in a 30-60-90 triangle. Students should be encouraged to understand this as a ratio rather than go straight to using a calculator. This is crucial to the conceptual understanding of trigonometry and is worth investing discussion time on now.

4 develops this understanding by showing that the sine and tangent ratios for 60° are constant in three 30-60-90 triangles.

5 confirms that the same properties hold for the sine, cosine and tangent ratios of a 30° angle.

What do you think? 12.4B

1 You may wish to explain the meaning of "cosine" as "sine of the complementary angle".

2 and **3** help students to develop an understanding of the sine, cosine and tangent ratios in a 45-45-90 triangle.

Consolidate – do you need more?

This mainly revisits the learning from the first part of the chapter, but question 5 looks again at the 30-60-90 triangle. Note that the values of the ratios are provided: students are not expected to learn these at this stage.

Stretch – can you deepen your learning?

1 tackles the potential misconception that any pair of shapes with equal angles are similar. Students need to appreciate that this is only true for triangles.

2 explores similar triangles from a different real-life context. Students may find it helpful to draw a sketch.

3 reinforces that congruence is "stronger" than similarity and that more information is needed to prove it.

4 introduces triangles where the side lengths are given as expressions rather than numbers.

5 leads students through the standard formal proof of the exact values of the trigonometric ratios for 30° and 60° angles and also demonstrates that $\sin x = \cos x$ is only true when $x = 45°$.

Reflect

Students reflect on the meaning of similarity and how this can be proved. Question 3 tests whether students understand how to work with ratios between the sides of the same triangle as well as ratios between the sides of similar triangles.

13.1 Direct proportion review

Small steps

- Solve problems with direct proportion (R)
- Explore direct proportion and conversion graphs (R)

Rationale

This chapter reviews and extends the work from Block 2 in Book 2. Since students have already explored direct proportion in an earlier block, the focus here should be on developing problem-solving skills and tackling multi-step problems. Students should be encouraged to explore different approaches, such as the use of proportion diagrams or double number lines.

Students met graphs showing direct proportion in Book 2, so there is an opportunity to consolidate this by making connections between this chapter and previous work on unit conversions, currency exchange and linear graphs. Students should be shown examples of graphs that do not show direct proportion for comparison (such as those which are linear but which do not go through the origin). Emphasise that a graph showing direct proportion must be a straight line through the origin.

Things to look out for	Interleaving opportunities
• Some students may be confused about when they need to divide and when they need to multiply. They should be encouraged to sense-check their answers, which will be especially important when exploring inverse proportion. • Pre-drawn axes can be provided for students who struggle to draw graphs. • Students may get confused when working with graphs that have different scales on the two axes.	• Work on direct proportion from Book 2 is reviewed and expanded upon. • Currency exchange from Chapter 8.4 is reviewed. • Students can revise plotting graphs and review their knowledge of linear graphs. • There are also opportunities to review unit conversions.

Models and representations

- The double number line, which has been extensively used in Books 1 and 2.
- Proportion diagrams which show the different relationships between variables can help students to explore different approaches.
- The graphs themselves are key representations in the second part of the chapter.

Behind the questions

Are you ready?

These questions provide a review of key numeracy skills as well as plotting and reading coordinates.

Practice 13.1A

1 to **3** require students to use proportional reasoning skills to move between the different quantities given. There is an opportunity for discussion around the efficiencies of different approaches.

4 provides a chance to move away from money by presenting the question in a different context. The numbers have been chosen so that students can use different methods rather than just the unitary approach. (For example, 75 g is $\frac{3}{4}$ of 100 g, so how can that help to work out the number of calories?)

5 and **6** involve familiar recipe contexts requiring the scaling of ingredients. In each question, part b offers additional problem-solving opportunities.

7 introduces the algebraic representation associated with direct proportion. This can be linked to the equations of graphs if desired.

What do you think? 13.1A

1 is a slightly more challenging proportion question which lends itself to proportion diagrams and exploring multipliers.

2 exposes students to a non-example of direct proportion to reiterate that the variables must increase or decrease at the same rate if they are to be directly proportional. Students could be asked to explore the graphical representation as an added challenge.

Practice 13.1B

1 invites students to review their graph drawing skills and to use their graphs to answer simple questions. Note that a tolerance should be allowed when reading from graphs.

2 provides a pre-drawn graph so that students can develop their analytical skills further.

3 provides an opportunity to review work from Block 8 by using a currency conversion graph. This question asks students to consider how to convert an amount which is beyond the range of the graph. This would be a good whole-class discussion task as several approaches are possible.

4 reinforces the definition of direct proportion. It is too easy to fall into the trap of "as one gets bigger so does the other". The non-example of temperature conversion from Celsius to Fahrenheit is used to emphasise this. Students need to appreciate that a graph showing direct proportion must go through the origin.

5 is a chance to develop students' language and reasoning skills as they explain why three of the four graphs do not represent direct proportion. Good mathematical language should be insisted upon, as always.

What do you think? 13.1B

1 challenges students to use two graphs to work out an exchange rate for themselves. It would be worth discussing their different strategies as there are a number of approaches that can be taken.

2 offers a chance to review other skills alongside direct proportion. Students will need to work out the volume of a cylinder, convert between metric units and use the given conversion graph to answer the question.

Consolidate – do you need more?

This section provides a chance to review all the key concepts from this chapter. It can be used to consolidate these skills if more practice is needed, or it can be used as a homework or revision task at a later date.

Stretch – can you deepen your learning?

1 is straightforward and provides a chance to review ratio and unit conversion in the same question.

2 helps to develop students' reasoning skills and their ability to explain their mathematical thoughts. Those who feel confident can be encouraged to use algebraic multiplier notation (e.g. if a is directly proportional to b then $a = kb$) to help them to explain their answer.

3 asks students what two graphs would look like, and this is an excellent opportunity to review linear graphs without needing to make accurate drawings (although they could, if they wish). For example, students could be asked "Which graph will be steeper?" and "How do you know?"

Reflect

These two questions remind students of the need for mathematical precision when defining concepts and reinforce the key features of direct proportion.

13.2 Investigating inverse proportion

Small steps

- Solve problems with inverse proportion
- Explore graphs of inverse relationships (H)

Rationale

In this chapter, students will explore and work with inversely proportional relationships. Using accessible contexts is key for enabling students to appreciate the relationship between the two variables. Within the classroom, students can be asked questions such as "If one student takes 30 minutes to tidy the classroom, how long would it take it two students? More time or less? By how much?" There are, of course, assumptions being made about the rate of work, and these should be discussed.

In the second part of the chapter, students can explore inverse proportion graphs. This is a Higher step and should only be approached if students are fully confident at working with inverse relationships.

Things to look out for

- Some students will mix up division and multiplication by "doing the same thing to both sides" rather than using inverse operations. Context can be very powerful here and students should be reminded to sense-check their answers.
- Since the graph of direct proportion is linear, students often assume that inverse proportion graphs are also linear but with a negative gradient – examples such as rectangles with a fixed area can be used to explain why this isn't the case.

Interleaving opportunities

- Plotting coordinates and reading values from accurate graphs can be reviewed.
- Work from Chapter 1.4 on non-linear graphs can be reviewed.
- There are opportunities to review working with time.
- Questions on area and perimeter are included in this chapter.

Models and representations

- Bar models can be used to show how an amount of work can be shared between different numbers of workers.
- Tables of values can be used to show the relationship between two inversely proportional variables (i.e. that their product is constant).

Behind the questions

Are you ready?

These questions remind students about inverse operations and the key numeracy skills they need to proceed. There is also a chance to review plotting and reading coordinates in preparation for the graph plotting section.

Practice 13.2A

1 and **2** are both scaffolded problems that involve inverse proportion. Students can use bar models or tables of values to help answer these. The worked example models an approach that students could adopt.

3 to **5** continue the same style of questions in a variety of contexts. Throughout, students should be reminded to sense-check their work. Although speed, distance and time have not been formally covered, students should appreciate that the faster something travels, the shorter the time it will take to travel a fixed distance. In question 4, there is no need for formal calculations or use of the formula.

6 slightly increases the level of difficulty as students need to work with hours and minutes. Common errors here (such as working with time as a decimal) can be addressed if necessary.

7 addresses a common misconception that any two variables that change in opposite directions represent inverse proportion. The concept of changing at the same rate is sometimes dropped from definitions of inverse proportion, so the necessity for this should be reinforced.

8 uses a table of value as the key representation. It also touches on the idea of the equation linking the two variables as being, for example, xy = constant.

9 asks students to identify direct and inverse proportion from tables of values and to consider similarities and differences between the two.

What do you think? 13.2A

1 adds a new level of difficulty by introducing a problem with three variables (number of workers, number of letters, and time) and a mixture of direct and inverse proportionality relationships. Students can be asked to consider how different pairs of the variables are related.

2 introduces a level of complexity in that the number of workers available changes over time, so they will need to work out how much work will be done with five workers and how long it will take to complete the project once all eight workers are available.

Practice 13.2B

1 and **2** give students a chance to practise their graph drawing and analysing skills in familiar contexts. Students should be encouraged to make links with the work covered in Chapter 1.4.

3 gives a pre-drawn graph and students practise their skills of reading accurately. Part b asks students to explore the equation of the graph (in the form $xy = 16$) and to consider how this is linked to the context of area of a parallelogram. They do not need to consider $y = \frac{16}{x}$ at this stage.

4 uses the context of Boyle's law as an example of an inverse relationship. Students do not need to understand the formalities but are asked to read from the graph and to come up with a possible equation for the graph.

5 addresses the misconception that the graph of inversely proportional quantities is somehow the "opposite" of direct proportion graphs.

What do you think? 13.2B

This challenges students to consider a scenario in which one variable increases as the other decreases but that doesn't represent inverse proportion, and to explain why this isn't the case. There is the chance to draw a graph and to consider how this example is mathematically, graphically and contextually different from the earlier ones.

Consolidate – do you need more?

This section provides a chance to review all the key concepts from this chapter. It could be used to consolidate skills if more practice is needed, or it could be used as a homework or revision task at a later date.

Stretch – can you deepen your learning?

1 is another example in which there are three variables to consider and students need to think about which are linked by direct proportion and which are linked by inverse proportion.

2 adds some new layers to familiar problems by considering the effects of varying the area and the number of workers as well as how much the project will cost.

3 looks much easier than it is, and many students will apply common misconceptions (for example, by finding the average time). Students could be asked to consider "How much of the room will be painted after one hour?" and to work ahead from there.

Reflect

These two questions ask students to reflect on the differences between direct and inverse proportion. Students can consider this in the context of a particular problem and/or use a graphical approach.

13.3 Ratio revisited

Small steps

- Solve ratio problems given the whole or part (R)
- Solve problems involving ratio and algebra (H)

Rationale

The first part of this chapter reviews and revises students' knowledge from Book 2. Since students have met and worked with ratio before, the emphasis here is on solving a variety of problems in order to hone students' problem-solving skills. Bar models are encouraged throughout as they will often make the course of action to take very clear. In the second part of the chapter, students can be extended by exploring problems which involve ratio and algebra. This is a Higher step and should only be tackled if students are confident with ratio and their algebraic manipulation skills. There is plenty of opportunity to extend students' algebraic thinking.

Things to look out for

- Students need to appreciate that the order in which the values in ratios are written is important.
- Students should be encouraged to read questions carefully so as to ascertain what information they have been given (the whole? the part? the difference between two parts?).
- Students commonly make the mistake of always adding the given numbers, even when they are not the parts that make up the whole. Bar models can really help to dispel this myth!

Interleaving opportunities

- Ratio from Block 1 in Book 2 is reviewed.
- Fractions and percentages are included within the practice questions.
- Students work with mixed units (such as grams and kilograms) and can review unit conversions.
- Ratios written in the form 1 : n and what this represents are reviewed.
- There are opportunities to review algebraic notation and solving equations.
- Geometry facts such as angles in polygons are used as contexts.

Models and representations

- Bar models can be used to represent the information provided in a question and to see how to proceed.
- Double number lines are helpful for showing multiplicative relationships.

Behind the questions

Are you ready?

These questions review the basics of ratio and fractions, and include equivalent fractions and converting between the two forms of notation.

Practice 13.3A

1 and **2** are straightforward ratio questions that remind students about equivalences as well as measure.

3 is a typical "sharing" question with bar models provided. This should help students to model the remaining questions in the exercise.

4 enables students to appreciate the value of bar models. They are given four questions with the same numbers and the same ratio but each represents a different scenario and needs to be modelled carefully.

5 is in a similar vein to question 4 as it encourages students to think carefully about what is being asked. It also adds working with money into the context.

6 is a good opportunity to discuss key misconceptions when moving between ratio and fraction notation. Students could be asked to think of some more statements of their own which are true or false.

7 reminds students to read the question carefully. Although there are three parts to the ratio, only one piece of information is given. They need to recognise that this actually represents two of the ratio parts.

8 reviews 1 : *n* notation in a familiar context and enables students to see the value of such notation. Here, they work out how much chocolate there is for one person, which makes comparison much easier.

9 uses perimeter as a context, and students need to take care not to just do "4 + 1" and then "50 ÷ 5" since this would mean sharing the perimeter across just two sides.

10 is more complex and requires problem-solving skills. Bar models can be very helpful here.

What do you think? 13.3A

1 reviews the definition of prime numbers and also develops students' language and reasoning skills as they need to explain carefully why the ratio will not simplify.

2 requires students to consider what the ratio tells them about the total number of counters.

3 is a typical problem involving the ratio of $a : b$ and the ratio of $b : c$ in which students need to think about common multiples in order to find the ratio of $a : b : c$.

Practice 13.3B

1 and **2** help students to develop intuition in moving from ratios to algebraic relationships via bar models. It's important for them to understand that the same relationship can be expressed in different ways.

3 builds on the two earlier questions and includes some substitution to warm up algebraic skills.

4 and **5** encourage students to set up and solve simple equations in order to find the value of x. Note that although question 4 can be answered by means of equivalent ratios (rather than with formal algebra), question 5 has been designed so that it cannot be. Therefore, question 4 should be used to model the approach needed.

6 and **7** both explore relationships between three abstract quantities using ratios, fractions and percentages. There are many approaches to these and students can take time to discuss their methods and strategies.

8 and **9** use familiar contexts to further develop algebraic skills and question 9 provides a chance to review work on surface area and volume from earlier in the year.

10 and **11** are typical "changing ratio" problems. Both are challenging and Example 4 can be used to model a suitable approach.

What do you think? 13.3B

1 is another changing ratio question but with the added difficulty that marbles are moved from one box to the other. Students need to set up an appropriate equation before solving it.

2 is much more challenging than it seems. The values of the parts in the first ratio are not the same as the values of the parts in the second, so students need to work out a relationship between the two. Some bar modelling will help students to picture the situation.

Consolidate – do you need more?

This section provides a chance to review all the key skills from this chapter. It can be used for further practice for those who need it, for homework or for revision at a later date.

Stretch – can you deepen your learning?

1 investigates what the ratio parts tell you about the whole – in this case, that it must be a multiple of 6.

2 is a working-backwards problem; students know the final amount that Sven receives and they need to undo each step to work out the original amount of money.

3 is another example of changing ratios and will give students a chance to develop their algebra skills.

4 is a problem for which there are many possible relationships between a and b.

5 extends the changing ratios problems by asking students to work out when a particular ratio will be arrived at.

Reflect

1 gives students a chance to create their own problems using the ratio 2 : 3 and 120 sweets. They could be encouraged to create more complex problems and to introduce fractions, percentages, etc.

2 is a chance to review moving from a ratio to an algebraic representation.

13.4 Best buys

Small steps

- Solve best buy problems

Rationale

Students will have explored best buys in Chapter 8.4. The focus then was on the unitary method as the main approach to problem solving. In this chapter, students are encouraged to explore other approaches, such as scaling up or down to get the price of the same quantity (not just limited to 1). They are encouraged to use different approaches so that they can get a better sense of when each method might be most effective.

Non-calculator methods and estimation can also be used as these apply more closely to real-life shopping without a calculator!

Things to look out for

- Finding the cost per item is a more natural approach but students should be exposed to amount per unit cost as well. Links should be made between the calculations (cost ÷ amount) and what this represents in the context (£ per ml).
- Students should appreciate that the best buy or best value may not be the same as the cheapest item. So although 1.5 kg of butter may be the best value, it may not be the cheapest or most useful size to purchase!

Interleaving opportunities

- Students will review work from Chapter 8.4.
- Students will revisit rates in Chapter 14.3, where connections can be made.
- Students will also review working with money.

Models and representations

- Bar models can be used to demonstrate the unitary method.
- Real-life offers and supermarket websites can be used to explore these ideas further.

Behind the questions

Are you ready?

These questions review working with money and rounding since these are necessary skills. They also remind students of the unitary approach.

Practice 13.4A

1 and **2** ask students to approach the problem in different ways. Students will find the cost of one unit, the quantity bought for £1 and also the cost of a scaled-up version. Each time, they should consider which method they prefer and why.

3 and **4** leave students to make their own choices about their preferred method. They could be asked to explain why they chose a particular approach rather than another.

5 deals with a common mistake arising from working with mixed units. The question includes litres and millilitres as well as pounds and pence. Students need to spot that Sven has made one correct conversion but forgotten the other. They should be reminded of the usefulness of sense-checking that their answers are appropriate.

6 introduces another dimension and theme from this chapter: that the best buy isn't always the most practical. If Darius wants 35 bars, it will be cheaper to buy 21 under one offer, 3 under another and one at full price.

7 is similar to question 6 in exploring best buys and real-life practicalities.

8 asks students to explore two different offers, working with different units to determine which is better value.

9 reminds students of the scaling-up approach. Ed is working out the cost of 60 cupcakes in each case. Discussion should be encouraged about why this is such an efficient method in this example.

What do you think? 13.4A

1 develops students' thinking and reasoning by giving them the calculations and asking them to consider what each calculation means in the context of the question.

2 reviews work on exchange rates from Chapter 8.4. Students need to choose a currency to work in, and then work out which box offers better value.

3 could form the basis of an investigation into which offer is best when buying different numbers of DVDs. Here, only 3 and 6 DVDs are purchased but students can create a table of values identifying the costs and looking for patterns.

Consolidate – do you need more?

This section provides a chance to review all the key skills from this chapter. It can be used to supplement Practice 13.4A, as a homework or as revision activity at a later date.

Stretch – can you deepen your learning?

1 asks students to compare two different offers and to decide which is better value. They could be asked to investigate whether the answer would change if a different number of bottles was being purchased.

2 gives the calculations for finding the best buy but leaves students to work out the size of the jars and the price of each one. This is another chance to explore the scaling-up method since, in this example, it makes the calculations much easier.

3 requires students to work with exchange rates and to convert between kilometres and miles before deciding which taxi trip offers better value for money.

Reflect

1 asks students to review the meaning of "best value" and why sometimes this isn't the cheapest or most practical option.

2 is a chance to create a poster or revision card to remind students of the different approaches to best-buy problems.

14.1 Speed, distance and time

Small steps

- Solve speed, distance and time problems without a calculator
- Solve speed, distance and time problems with a calculator
- Use distance–time graphs

Rationale

As an introduction to the block that studies rates, students look at speed as this is likely to be familiar from their everyday experience. As well as informal methods, the formula speed = $\frac{\text{distance}}{\text{time}}$ is used throughout, and students are expected to be able to substitute the values they know and rearrange the formula to find the value they want. This reinforces their earlier skills rather than using potentially unreliable "formula triangles". Students also explore distance–time graphs, which links this topic to the gradient of a straight line.

Things to look out for

- Not understanding that "per" means "for every". Discuss, for example, that a speed of 60 mph means travelling 60 miles in 1 hour, 120 miles in 2 hours, etc.
- Thinking that the sloping parts of a distance–time graph represent motion uphill and downhill. Use a step-by-step approach to the construction of graphs and discuss this misconception.
- Treating time as a decimal quantity, for example thinking that 30 minutes is 0.3 hours. Address this explicitly when comparing calculator and non-calculator methods.

Interleaving opportunities

- Ratio
- Direct and inverse proportion
- Forming and solving equations
- Gradient of a line segment
- Reading scales
- Choosing and using scales
- The concepts in this chapter are developed further in the rest of the block.

Models and representations

- Distance–time graphs are a key model for representing journeys with several stages and need careful explanation as most graphs that students have met up to now will have had a single section only. In particular, you will need to discuss the meaning of "flat" sections, i.e. that the object is stationary or the speed is zero (as the rate of change is zero), and that the greater the gradient (positive or negative), the greater the speed.
- The double number line (see question 3 in Practice 14.1A) helps to make sense of the "per" in, for example, "kilometres per hour" and makes clear the links with ratio and direct proportion.

Behind the questions

Are you ready?

Question 1 looks at solving equations of the forms that will be met in this chapter and question 2 develops this into rearranging formulae as an alternative approach. Question 3 looks at conversion between hours and minutes and question 4 practises rounding when using a calculator.

Practice 14.1A

1 and **2** explicitly look at the meaning of "per" and show that informal methods based on the understanding of speed as a rate of change can be used to answer some questions without the need for formulae.

3 uses a double number line representation, with the speed being the constant multiplier between the two horizontal lines.

4 allows students to practise using the formula but without the need for a calculator. Alternatively, the answer can be found using reasoning and an understanding of speed as a rate. Part b is a useful starting point for a discussion around the fact that speeds are generally unlikely to be constant due to traffic conditions, other events, etc.

5 and **6** deepen students' understanding by looking at different units of speed.

7 uses data about athletics. Students could also look at other races, or their own data, and again appreciate that speed will not be constant over the entire course because of the initial time needed to accelerate, etc.

8 develops the use of the formula to find times given speeds.

9 tackles the misconception of treating time as a decimal.

What do you think? 14.1A

1 provides the starting point for a discussion that compares informal and formal methods of dealing with fractions of an hour.

2 looks at unit conversions. Students may need to be reminded that 5 miles is about 8 km.

3 looks at a more complex time calculation that can be solved with the support of a number line.

4 invites further investigation of real, relevant calculations where the numbers are likely to be more challenging.

Practice 14.1B

1 This simple graph of a journey in one direction introduces students to the features of distance–time graphs without unnecessary complications.

2 asks students to compare two graphs, again without a change in direction.

3 provides the opportunity for students to draw their own graphs, this time with a return journey involved. You may wish to do this as a class, or refer students to Example 2 as a guide.

4 develops the idea of average speed without the support of a given graph. Students could draw a rough sketch for support.

5 to **7** develop this further, with different information for different parts of the journeys. Encourage students to consider whether their answers make sense.

8 is another comparison question, where students are required to sketch their own graphs.

What do you think? 14.1B

1 tests students' understanding of gradients and directions. They could also create their own travel graphs to discuss.

2 introduces more realistic graphs and the idea of acceleration. This is a good point at which to discuss the limitations of the mathematical models used in the graphs earlier in the chapter.

Consolidate – do you need more?

This revises all the key points covered in the chapter and can be used for further practice, either now or later.

Stretch – can you deepen your learning?

1 links speed with standard form. You could extend this question by looking at the time it takes for light to travel between planets, stars, etc.

2 looks in more detail at the idea of representing both acceleration and constant speed on a graph. This could form the basis of a whole-class discussion.

3 tackles the common misconception about average speed in an algebraic form. This can be easily extended to produce expressions for travelling for twice as long at a particular speed, for example.

4 links back to Block 13 and direct and inverse proportion. You could also discuss graphs.

Reflect

In the first question, students are encouraged to think about the meaning of the formula for speed and how it can be expressed in different ways to help to work out the different variables. The second question focuses on understanding of graphs and is best approached by means of a labelled illustration.

14.2 Density and units

Small steps

- Solve problems with density, mass and volume
- Convert compound units (H)

Rationale

This builds on Chapter 14.1 by looking at another rate that shares the same structure as speed, namely, density. Again, formula triangles are avoided in favour of double number lines, as double number lines rely on understanding. Students will continue to practise their algebra skills by rearranging the formula when necessary. As students have now seen the idea of a rate as "this per that" in two contexts, you can either move them on to the more general rates in Chapter 14.3 or, if appropriate, tackle the Higher step of converting compound units that forms the second part of this chapter.

Things to look out for	Interleaving opportunities
• Not understanding the notion of density. Explain this by comparing some real objects, such as cotton wool and bricks, that provide a clear contrast. • Confusion with units. Students should be confident at identifying the units of mass and of volume, so the focus should be on understanding density as "mass per volume" and making the link with "distance per time".	• Ratio • Direct proportion • Forming and solving equations • Units and unit conversions • Names of 3-D shapes, and their volumes • Density as a rate can be compared with speed from Chapter 14.1; come back to this in Chapter 14.3.

Models and representations

- Using real objects in the classroom can help students to make sense of the notion of density. You could challenge students to think about their own possessions, e.g. the mass of a book or mobile phone compared with their volumes.
- The double number line, linked to the previous chapter, continues to help students to understand "per". You can look at both "per g" and "per kg" and discuss which is more appropriate. This is linked to the Higher step of converting units that comes later in the chapter. The double number line also makes clear the links with ratio and direct proportion, with density being the constant multiplier.

Behind the questions

Are you ready?

Questions 1 and 2are simple reminders of units of mass and how to find a simple volume. Make sure that students use the correct units in their answers to question 2. Question 3 checks that students can solve equations of the form $a = \frac{b}{c}$ to find each of a, b and c. Question 4 checks understanding of rates and asks students to choose the correct units in the context they met in the previous chapter, i.e. speed.

Practice 14.2A

1 introduces using the formula density = $\frac{\text{mass}}{\text{volume}}$. In the interests of realism, the numbers have not been simplified. You may need to check rounding and also that the correct units, grams per cm^3, are given with students' answers. You could discuss how students can identify the units from the question.

2 develops this by looking at a different possible unit for density, in the context of a liquid.

3 still requires use of the density formula for basic fluency calculations but also introduces an element of reasoning with "more than double". Compare students' solutions and how they present this simple proof, and discuss what is needed to make the solution clear.

4 and **5** involve using the density formula to find mass and volume. Students need to make sense of the given units of density to give the correct units in their answers. This may need checking and discussing.

6 interleaves finding the volume of a triangular prism.

7 tests understanding further by requiring careful analysis of what information is given and what needs to be found. Encourage students to look at the units to support their thinking.

8 Students need to pay attention to the units, and the question requires some comparison of decimals.

What do you think? 14.2A

1 makes explicit the link with speed and invites discussion of the related formula for pressure.

2 is a reminder that the terms "mass" and "weight" are not interchangeable.

3 looks at combining densities, which you can link to average speed. This is developed further in the "Stretch" exercise, which involves calculation as well as reasoning.

Practice 14.2B

1 models a step-by-step approach to converting units, by looking at one unit at a time, in the now familiar context of speed.

2 develops this with a different starting speed. You may wish to discuss the idea of average rates again here as, in the context of a bus, the distances covered in the given time intervals are extremely unlikely to be constant.

3 and **4** The scaffolding is removed. If students struggle, encourage them to reflect on the step-by-step models given earlier.

5 requires a comparison. Students can compare converting the first speed to km/h with converting the second speed to m/s as both approaches are equally valid.

6 challenges thinking skills by looking at an unfamiliar unit where the conversion factor is not explicitly given. However, the actual number of calculations needed is small.

7 also looks at density but with familiar units.

8 uses imperial units. Although students are likely to be familiar with the names, they are often unaware of the conversions. These are, therefore, provided.

What do you think? 14.2B

1 involves converting between metric and imperial units. You could suggest drawing a double number line for support.

2 challenges students to simplify the process seen at the start of the exercise by generalising to a single multiplier. You could also look at converting the other way round and make the link with a reciprocal.

3 combines converting between larger and smaller units with metric and imperial conversion.

Consolidate – do you need more?

The main focus is consolidation of the work on density, but some further practice on the Higher step is also included.

Stretch – can you deepen your learning?

1 links to average speed as students need to find total mass and total volume from different given information.

2 requires the same skills, but adds challenge by giving the resulting density and requiring students to work backwards. Careful choice of the correct calculation is needed.

3 is a typical GCSE-style question that links density to finding the volume of a compound shape and converting units. Although each step is not difficult in itself, this needs an organised approach. It will be useful to compare and discuss different ways of presenting the solutions to multi-step problems.

Reflect

This checks students' understanding of the concept of density and use of the formula density = $\frac{\text{mass}}{\text{volume}}$.

14.3 Rates and graphs

Small steps

- Solve flow problems and their graphs
- Calculate with rates of change and their units

Rationale

This final chapter in the block builds on the previous two by exploring rates in a wide variety of contexts. Students should by now have a secure grasp of the concept of "this per that" and working out different aspects in a problem, i.e. each of the variables and the rate itself. In this general form, there is no formula provided, so "making sense" of the situation is crucial. Conversion between currencies, as studied in Books 1 and 2, is included as an example to make the link with the phrase "exchange rate". This is also an opportunity to look again at graphs, including conversion graphs, and to make the link with gradients and average rates.

Things to look out for	Interleaving opportunities
• Not being sure whether to multiply or divide. Encourage students to think about the meaning of the rate or variables given and about the units needed in the answer to decide which calculation needs to be performed. • Difficulty converting between hours and minutes. Model this and reinforce the fact that 3.56 hours, say, does not mean 3 hours and 56 minutes.	• This builds on the learning of Chapters 14.1 and 14.2 to explore general rates. You can link back to this previous learning to support students. • Ratio • Direct and inverse proportion • Forming and solving equations • Unit pricing • Exchange rates • Calculating and interpreting the gradient of a line segment

Models and representations

- The double number line is now familiar for modelling the "for every" idea that rates of "this per that" are built on. This can be used to support the solution of many of the questions in the chapter.
- A graph is also included, as this is the focus of the second part of the chapter. Students need to appreciate the link between gradient and rate. You might wish to discuss which axis it is more appropriate to use to represent each variable.

Behind the questions

Are you ready?

Question 1 is a reminder of exchange rates and can be used to start a discussion about "dollars per pound" and "pounds per dollar" that will support learning later in the chapter. Question 2 revisits speed and question 3 is a reminder of how to find a gradient from a graph, which will be essential for understanding the graphical rates later in the chapter.

Practice 14.3A

1 "Words per minute" is another rate that students should be able to relate to. A double number line can be used for support. This is another chance to discuss "average" rates by considering different sorts of text and passages with longer/shorter words.

2 You could draw a sketch graph to illustrate this and make the link with gradient as an introduction to the second part of the chapter.

3 This question involves finding a rate and then using it in a calculation. A double number line can be used to represent this. The fact that the rate is not likely to be constant at all hours of the day leads to another useful discussion about the accuracy of the model.

4 looks at another context. It might be worth discussing the apparent contradiction of mixing imperial and metric units, but the rate is valid if the units are clear.

5 links rates to unit pricing/best buys, with the simple rate of "cost per litre".

6 links to standard form.

7 and **8** link to direct and inverse proportion.

What do you think? 14.3A

1 looks at a common misconception when trying to find an average rate. Students could be guided to find the correct answer by considering a room of area 15 m^2 (or another common multiple of 3 and 5) and working out the individual rates and then the joint rate by adding these. You could extend this by generalising it to any given number of hours.

2 interleaves error intervals and the effect this has on the calculations of rates. This links to upper and lower bounds at KS4.

3 uses algebra to generalise rates and form expressions and equations.

Practice 14.3B

1 Students need to make the link between the gradient and rate of flow. They then use this rate to solve simple problems of a familiar style for this block.

2 reverses this by asking students to draw a graph to show a given rate. Again, this is explicitly linked to the gradient. The last part compares two rates; the second rate could be added to the graph if you wish.

3 compares a set of graphs that represent similar situations. You may wish to make students aware that the final graph has time in seconds but that the others have time in minutes.

4 links back to density, with the gradient of the graph being the rate of "mass per volume". You could use this, and previous graphs, to compare graphs and double number lines as forms of representation.

5 explores interpreting gradients as rates of "something per something".

6 looks at rates that change over time and links with average rates. You can compare this with average speed and mixed density problems.

7 provides a chance to discuss rates that are not constant.

What do you think? 14.3B

1 explores the misconception concerning average rates. You could link this with average gradient or average speed.

2 builds on the idea of non-constant rates by thinking about the shapes of the graphs. You can extend this by looking at other 3-D shapes.

Consolidate – do you need more?

This exercise provides further practice for use in class or as homework.

Stretch – can you deepen your learning?

1 looks at multiple currency exchanges. You could take this further by looking at current exchange rates for various currencies.

2 links to the volume of a cylinder.

3 is slightly more complex and could be the starting point of an investigation about how the percentage change affects the rates and times by considering the different types of proportion involved.

4 is a GCSE-style question that links using rates and graphical representation. Students may need help to present the multi-step solution logically. You can compare this with the similar question in Chapter 14.2 (see question 3 in the "Stretch" exercise).

Reflect

This encourages students to make links between all aspects of the learning in this block.

15.1 Probability of one event

Small steps

- Calculate single event probabilities (R)
- Understand relative frequency
- Calculate expected outcomes

Rationale

This chapter revisits finding the probability of a single event, which was covered in Book 1, and it also introduces the idea of relative frequency and expected outcomes. The first part of the chapter focuses on the probabilities of single events. Students should be able to work confidently with probabilities that are expressed as fractions, decimals or percentages and should know that a ratio is not an appropriate way to express a probability. In the second section, students should understand that expected outcomes can be based on experimental or theoretical probability, whilst relative frequency is based on experimental data.

Things to look out for

- If there are two possible choices, students may incorrectly give the probability of each choice as $\frac{1}{2}$ without considering the actual number of possible outcomes. For example, if there are two different colours of sweets in a bag, they may think that the probability of selecting each colour is $\frac{1}{2}$ regardless of how many there are of each colour.
- Students may be confused by the word "frequency" and think that "relative frequency" and "expected outcomes" are the same. Consistent use of correct language can help to prevent this.

Interleaving opportunities

- Students will be given many opportunities to work with fractions, decimals and percentages throughout the chapter.
- Types of numbers, e.g. prime and square numbers, can be revisited through probability questions.
- Algebraic methods can be applied in the context of probability to enable students to consolidate their algebra skills.

Models and representations

- A probability scale can be used to show that probabilities are given on a scale from 0 to 1 and also that probabilities sum to 1.

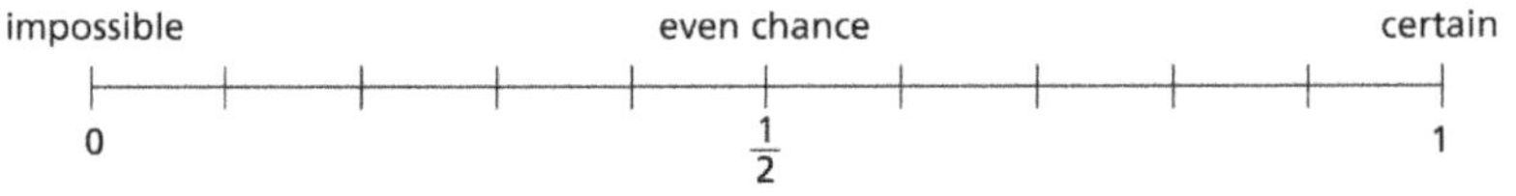

Behind the questions

Are you ready?

Questions 1 and 2 ensure that students understand what is meant by an outcome and are able to list these for common events. Questions 3 and 4 revisit previous learning on fractions.

Practice 15.1A

1 In part c, students should be able to articulate that because all possible outcomes are covered, the sum of the probabilities is 1.

2 Parts d and e link to the idea of something being impossible or certain and how these are expressed as probabilities. You could have an interesting discussion around why the values in parts a, b and c don't sum to 1: students should recognise that not all of the possible outcomes are covered.

3 Encourage students to notice that their answers to parts a, b and c sum to 1, as do the answers to parts c and d. They could compare their methods of calculating the answer to part d.

5 Students should recognise that the probabilities being calculated in parts d and e are the same because if the counter is not blue then it must be either orange or yellow.

6 provides opportunities to reiterate the importance of the selection being random. Students could discuss why, if the selection wasn't random, these probabilities are unlikely to reflect the decision-making process.

7 to **10** are designed to reinforce the idea that probabilities sum to 1. In question 7, students know the total number of students and so could choose to first calculate the number of boys before writing the probability. You could use this question to remind students that probabilities of all possible outcomes must sum to 1. Questions 8 to 10 remind students that probabilities can be expressed as fractions, decimals or percentages. Question 9 could form the starting point for a discussion about the effect of having other colours of sweets, not just orange and green, in the bag.

What do you think? 15.1A

Question 1 addresses the misconception that all outcomes are equally likely. In question 2, students might be tempted to go straight to finding a common denominator to compare the fractions but they could instead consider whether each fraction is greater than or less than $\frac{1}{2}$. In question 3, notice that the probabilities given sum to 1 and so could cover all possible options; however, you could have an interesting discussion around which numbers satisfy more than one of these conditions, and students could be challenged to consider how many answers they can find. Question 4 interleaves previous learning on ratio and students should be encouraged to consider what they think the common mistakes are when interpreting a probability from a ratio.

Practice 15.1B

1 and **2** Students use the idea of theoretical probability in familiar contexts.

3 reinforces the idea that it is not just the different outcomes that need to be considered but also their probabilities.

4 builds on the ideas covered in the previous three questions but presents some of the probabilities as decimals and percentages. Students could use their knowledge of fractions, decimals and percentage equivalence: e.g. 0.37 is equivalent to $\frac{37}{100}$ so you would expect 37 out of every 100 rolls to be green.

5 and **6** move on from expected frequency to relative frequency. Students should understand that relative frequency could change with every trial and is different from the theoretical probabilities they've seen previously.

What do you think? 15.1B

These three questions provide opportunities for students to think more deeply about relative frequency and expected outcomes. Questions 2 and 3 highlight the importance of the number of trials and how increasing this improves the accuracy of the data collected.

Consolidate – do you need more?

These questions provide opportunities for students to consolidate their understanding of the learning from this chapter. If further practice is needed, students could work in pairs to create similar questions of their own.

Stretch – can you deepen your learning?

1 You could suggest to students that they first consider what calculation they would do if the probability was 0.2 or $\frac{1}{7}$, say, before forming an expression in terms of x.

2 Similarly, you could suggest that students assign values to a and p. Students could be challenged to consider the limitations on a and p and generalise that a must be an integer such that $a < p$. They could also be asked to find an expression for P(not red).

3 In part b, students should recognise that the expressions for parts iv and v are equivalent as they represent the same probabilities.

4 and **5** Students could be encouraged to draw tables to structure their thinking.

6 interleaves previous work on algebraic manipulation in the context of probability. Note that students do not need to solve the quadratic as they aren't asked to find the value of x.

Reflect

This encourages students to reflect on what they have learnt in this chapter and to explain it in their own words.

15.2 More than one event

Small steps

- Calculate with independent events
- Use diagrams to work out probabilities

Rationale

In this chapter, students are introduced to the probability of independent events, which is sometimes referred to as the "and" rule for probability. Students may have previously seen the product rule in Book 2 and should use this, along with sample space diagrams, to support their understanding of independent events. The second part of the chapter revisits finding probabilities from diagrams; again, this will be familiar from Book 2.

Things to look out for

- Some students may try to generalise to derive a rule too quickly, which may result in an incorrect method for finding the probability of independent events. If this happens, students should be encouraged to draw sample space diagrams to identify their mistakes.
- A common mistake is to use the wrong denominator or "whole" when finding probabilities. Emphasise the importance of reading each question very carefully to identify the correct value.

Interleaving opportunities

- Knowledge of fractions, decimals and percentages will be required throughout.
- Students will have opportunities to revisit the key learning from the previous chapter.
- Addition and subtraction are used throughout when students are completing diagrams in order to find probabilities.
- Algebraic notation can be revisited in the context of probability.

Models and representations

- Students should be familiar with both two-way tables and sample space diagrams from their learning in Blocks 5 and 6 of Book 2.

Behind the questions

Are you ready?

These four questions remind students of some essential skills that will be needed throughout this chapter. Questions 1 to 3 revisit learning from the previous chapter on the probabilities of single events whilst question 4 recaps multiplying fractions, which was covered in detail in Book 2.

Practice 15.2A

1 is designed to introduce students to calculating probabilities of independent events and to encourage them to make connections. In part f, students could use both methods of calculating each probability and then generalise their findings.

2 Students could be prompted to explain the importance of the selection being random, and why this probability might not be the actual probability of that outfit because of other influences.

3 Students could draw a sample space diagram to support their thinking. However, watch out for students setting up their sample space incorrectly, e.g. only having one 3 listed in the outcomes for the first spinner.

4 looks at the probabilities of independent events in a more abstract way. Notice that, in part a, both fractions are unit fractions, and the answer doesn't need to be simplified, whereas in part b, the fractions are non-unit and the answer will need to be given in its simplest form to match the one shown in the question.

5 This question requires students to formally discuss independent events in more detail and to explain why their probabilities do not necessarily sum to 1.

What do you think? 15.2A

For question 1, a sample space diagram could help students to recognise the mistake that has been made and will highlight the importance of considering the different orders. Students should then use what they have learnt to help them to calculate the probability in question 2. Question 3 addresses the common error of drawing a sample space diagram representing the possible outcomes but not considering their probabilities.

Practice 15.2B

1 Notice that the two-way table is already complete so the focus is on extracting the probabilities from this information. Watch out for students giving their probabilities out of 320 in part b.

2 Here the students' focus is on finding probabilities from a frequency tree. Again, students should carefully consider what the denominator of their fraction will be and why.

3 Watch out for students thinking that only 42 students play football or only 29 play netball and not fully understanding what is represented by the 30 in the intersection.

4 Students now need to complete the two-way table themselves before extracting the required probabilities. An interesting topic for discussion could be which probabilities can be found without completing the two-way table and which require it to be completed.

5 In part b, students might be unsure of what the denominator of their fraction should be. Encourage them to read the start of the question again and draw their attention to the 300, which is given in the first line.

6 In part a, students should notice that the numbers given don't sum to 100 and should be able to explain where the other people would be represented on the Venn diagram. You could suggest that they copy the diagram and fill in the numbers to help them to find the required probabilities.

What do you think? 15.2B

Rather than having a partially or fully completed diagram, here students are required to draw their own from scratch. Students should recognise that there isn't necessarily only one way to construct these diagrams; however, the probabilities extracted should be the same. In question 3, students need to decide which diagram to draw and should be encouraged to explain their choice.

Consolidate – do you need more?

These questions provide further practice at calculating probabilities of more than one event and finding probabilities from diagrams. If more questions are needed, students could work in pairs to create similar ones of their own or be asked to calculate other probabilities from the information given in these questions.

Stretch – can you deepen your learning?

1 Students should be encouraged to explain why these events are independent. They could be challenged to find the probability that Ed does not pass his driving test first time but scores a goal in his football match on Saturday.

2 Some students might try to just list as much information as possible without organising it in a clear and methodical way. Encourage them to use different types of diagram and discuss which is most useful.

3 Students use knowledge of division of fractions to find that P(B) would have to be greater than 1, which isn't possible as probabilities must be between 0 and 1.

4 Students should be encouraged to give their answers in their simplest form by using rules of indices, which were covered in Book 2.

5 Students could draw pictures here to support their thinking or use concrete manipulatives to represent the situation. This question leads into the next chapter, which will look at probability tree diagrams.

Reflect

These two questions invite students to explain their learning in their own words. For question 1, they could be asked to give examples of two events which are independent and two which are not to demonstrate their understanding. Question 2 links back to the idea covered in parts a and b of question 1 in Practice 15.2B. Refer students back to this question to help them with their explanations.

15.3 Probabilities and tree diagrams (H)

Small steps

- Use tree diagrams (H)
- Use tree diagrams to solve "without replacement" problems (H)

Rationale

In this chapter, students learn how to use tree diagrams to help them to find probabilities. The first part of the chapter focuses on examples involving independent events ("with replacement") whilst the second part introduces the idea of dependent events ("without replacement"). When calculating the probability of each outcome, or "across branches", links should be made back to the previous chapter so that students understand that they are calculating probabilities of independent events. Throughout this chapter, if necessary, you could provide students with partially completed probability tree diagrams. As they become more confident, you could reduce the amount of scaffolding. Students should be encouraged to list all of the possible outcomes to the right of their tree diagrams for each question as this will help them to interpret the probabilities.

Things to look out for

- Students may try to perform too many steps at once when labelling probabilities: for example, if the question involves two pens being selected, they may subtract 2 from the numerator/ denominator of the fraction in one step.
- Incorrectly labelling the probabilities on the second set of branches when working with dependent events.
- Numerical errors can occur, particularly when probabilities are given as decimals rather than fractions. Encourage students to use the fact that probabilities sum to 1 as a method of checking their answers.

Interleaving opportunities

- This chapter will naturally interleave and build upon all the learning from earlier in this block. Students will need to use their knowledge of probabilities of single events and probabilities of independent events throughout.
- Multiplication of fractions is required when calculating probabilities from tree diagrams.
- Some questions require students to use their understanding of algebraic notation in the context of probability.

Models and representations

- The tree diagram is a new representation and will need step-by-step modelling a few times to help students understand how it works. You could discuss the similarity in structure with that of frequency trees students used in Books 1 and 2.

Behind the questions

Are you ready?

Question 1 revisits basic probability from earlier in the block whilst questions 2 and 3 consider the ideas of replacement and without replacement respectively, but not in the context of probability. It is essential that students appreciate the difference. This provides a good starting point for a discussion point about how this changes the situation and how the wording in the question relates to this. The use of concrete manipulatives or pictorial representations can be used to support understanding.

Practice 15.3A

1 Students should be encouraged to label their branches "odd" and "even" rather than with the numbers 1 to 6. Although both methods will lead to the same probabilities, students should recognise that it is more efficient to draw a probability tree diagram that has four branches than one that has 36.

2 In part c, students should recognise that the probabilities sum to 1 because this covers all possible outcomes: either Chloe will win the game or she will not win the game.

3 The use of real counters, or a pictorial representation of them, can support students with this, if necessary. This question follows a similar structure to that of the worked example before the questions, so students can refer back to this if they need guidance.

4 In part b, students could be encouraged to reflect on the possible methods: they could either calculate P(PP) + P(PF) + P(FP) or 1 – P(FF).

6 Here the probabilities are given as decimals. Students may prefer to use their knowledge of place value to rewrite these as fractions.

What do you think? 15.3A

Question 1 highlights why simplified fractions are not necessary when calculating with probabilities. Although the probabilities calculated are correct, simplification of the fractions adds extra steps to the process when these are used in later calculations. Question 2 addresses the common misconception of adding all of the probabilities in a given column. Students should realise that each pair of branches needs to sum to 1 and be able to explain why.

Practice 15.3B

1 Students should be encouraged to reflect on the similarities and differences between this question and question 3 in Practice 15.3A. The context is similar to that of the worked example, so students could refer back to that for guidance if needed.

2 You could have an interesting discussion around whether these probabilities are likely to be accurate in the context of the question as people's choice of sandwich tends to be not random.

3 Part b i (the probability that there are no red pens left) might confuse students as it is slightly different from the questions they've seen previously. They should think carefully about which of the outcomes on their probability tree diagram represents the one required.

6 Here the question is presented using slightly different wording, and students should consider carefully which branch of their tree diagram represents each situation when labelling it. It might be useful to list the outcomes to the right before labelling the probabilities on the second set of branches.

What do you think? 15.3B

Question 1 explores the common error whereby students try to take too many steps at once using a probability tree diagram. If this happens, remind students that the probability tree diagram is a way of modelling the possible outcomes of the situation and that they need to consider each sweet, one at a time, in order to do this accurately. Question 2 requires students to work in reverse; they should be encouraged to label as much information as possible on their tree diagrams to help them to find the required probability.

Consolidate – do you need more?

These questions allow students to consolidate their understanding of finding probabilities using tree diagrams. Notice that the probability trees are drawn to support students and the amount of scaffolding is reduced slightly as they progress though the exercise.

Stretch – can you deepen your learning?

1 combines learning from this chapter with that of the previous chapter in a problem-solving context. Students should be able to explain why they can only estimate the number of prize winners and hence only predict the amount of profit/loss and are not able to calculate this accurately.

2 If students find this question difficult, they might need prompting to recognise that the sum of the three probabilities given must be equal to 1 as they are the only possible outcomes on the spinner. They can then form and solve an equation to find numerical values for the probabilities.

4 Questions like this can appear difficult to students and they often don't know how to get started. You could ask, for example, "What if p was equal to 17, then what would you do?" to guide them towards forming expressions for the given probabilities. At this stage, students do not need to be able to solve the quadratic equations formed – they just set them up.

Reflect

Encourage students to use annotated diagrams to support their reflections of the key learning points here.

16.1 Further graphs

Small steps

- Draw and interpret quadratic graphs
- Interpret graphs, including reciprocal and piece-wise

Rationale

Students have developed their knowledge of how to plot and read from linear graphs throughout the series, and this chapter extends their skills to working with other graphs. The initial focus is on quadratic graphs, where students learn about the shape and symmetry of parabolas. Note that some students may have looked at simple quadratics as a Higher step in Chapter 4.6 of Book 2. They then move on to consider other graphs, including other curves and piece-wise graphs.

Things to look out for

- Difficulty with directed number substitution. Revise this in a starter and/or use calculators for complex expressions. You may need to remind students how to square negative numbers correctly using a calculator.
- Students may need support when reading scales on graphs and choosing scales for their own graphs. Make the link with number lines and intervals explicit.
- Joining points with line segments instead of a smooth curve. Use graphing software to model the correct appearance and compare this with graphs drawn with line segments to show that these are incorrect.
- Some students will find it difficult to draw a smooth curve. Model this, and allow time for plenty of practice.
- Confusion over the endpoints in piece-wise graphs. Discuss the meaning of the filled and hollow circles.

Interleaving opportunities

- Linearity and non-linearity, linking to sequences material in Books 1 and 2 and/or the review in Block 17
- Using algebraic expressions to complete a table of values
- Directed number
- Inequalities and interval notation
- This chapter links with representing inequalities in Chapter 16.3.

Models and representations

- Students need to be confident at reading and using coordinate grids with different scales, including those with different scales on the two axes. You can use a table of values to illustrate why this is necessary for graphs such as $y = x^2$ and $y = x^3$, where the changes in y are much greater than those in x.
- Graphing software is really useful for students to quickly see what's the same and what's different about the graphs as you change the coefficients of the terms. You could model this or you could set it as an exercise for the students. Compare a variety of graphs to illustrate the shape and symmetry of the parabolas.

Behind the questions

Are you ready?

Question 1 reminds students of how to fill in a table of values for a linear graph and question 2 extends this to look at a graph with a negative gradient. Be aware that some students will have difficulty working with directed numbers; it is likely that any errors made when drawing linear graphs will be repeated when students move on to the more complex graphs. Question 3 progresses to using directed numbers in expressions involving x^2 and may indicate the need for revision/calculator support.

Practice 16.1A

1 Students will either need to line up a ruler on the textbook or have a copy of the graph to draw guide lines on. They will need to look carefully at which variable is given and how many answers to expect. This is all covered in the first worked example.

2 and **3** involve simple quadratic plots for students to practise drawing curves. Time and patience are needed to develop this skill!

4 looks at three-term equations, so students may need support when substituting values. They should notice that the extra term does not affect the parabolic shape of the curve.

5 requires students to create their own tables of values.

6 looks at a common error when the vertex of the curve is not a value in the table. Students' knowledge of the symmetry of the curves should help them to establish where the turning point should be.

What do you think? 16.1A

1 extends the topic by looking at negative parabolas.

2 and **3** require students to recognise types of graph from given equations. The reciprocal graph is covered in the next section.

Practice 16.1B

1 and **2** explore reciprocal graphs. This is the first time that students will have met discontinuous graphs, so this will need to be discussed; include the fact that division by zero is undefined.

3 gives an example of an exponential curve. You could make the link with real-life graphs in your teaching here.

4 and **5** look at another type of discontinuous graph – piece-wise graphs – where the sections are line segments. You may wish to discuss the end points before students tackle the questions.

6 Explain to students that they have already met a different form of piece-wise graph when studying distance–time graphs.

What do you think? 16.1B

1 Students draw their own piece-wise graph from a table of information.

2 and **3** link to equations of straight lines as students are asked to draw their own piece-wise graphs from equations.

Consolidate – do you need more?

This provides further practice at drawing quadratic, reciprocal and piece-wise graphs.

Stretch – can you deepen your learning?

1 looks at equations of lines of symmetry. This could be checked using software.

2 builds on question 1 and gives students the opportunity to generalise the equation of the line of symmetry of a quadratic curve.

3 looks at the effect of constants in different places in the equation of a reciprocal curve.

You can extend this "Stretch" section to look at cubic and other graphs if time allows.

Reflect

Students reflect on their learning about quadratic graphs by noting the similarities and differences between these and linear graphs and their equations. They also reflect on piece-wise graphs.

16.2 Simultaneous equations (H)

Small steps
• Investigate graphs of simultaneous equations (H)

Rationale

Students use their knowledge of straight-line graphs to solve simultaneous equations graphically. Using a table of values and considering where the graphs cross the axes (sometimes called the "cover-up method") are both considered. Although an algebraic approach is not required at KS3, a few examples that can be solved using bar models are included for interest.

Things to look out for	**Interleaving opportunities**
• Sign errors when finding the values where the graphs cross the axes. Encourage students to use a test point to check. • Students should be encouraged to check whether their answers are correct by substituting them into the original equations as the graphical method is not entirely reliable, especially for non-integer solutions.	• Solving equations • Substitution into expressions • Straight-line graphs

Models and representations

- Understanding coordinate grids is essential, particularly if the scales are non-unitary. Some geoboards have axes labelled and these could be useful models for students.
- Straws can be used to model the number of possible intersections of straight lines, e.g. two lines can meet once (the straws cross or will cross if extended), will never meet (the straws are parallel) or meet an infinite number of times (the straws lie on top of each other, analogous to $x + y = 10$ and $2x + 2y = 20$, say).
- Graphing software can also be used to illustrate when/if pairs of straight lines will cross and to read off the point of intersection.

Behind the questions

Are you ready?

Question 1 checks that students are familiar with the equations of straight lines parallel to the axes. Question 2 revises drawing a linear graph using a table of values, and question 3 revises parallel lines.

Practice 16.2A

1 The idea of a point of intersection is introduced by looking at the coordinates of the point where a line parallel to the x-axis meets a line parallel to the y-axis.

2 develops this by drawing another pair of lines of the form $x = \ldots$ and $y = \ldots$.

3 The scaffolding of being given the drawing is removed as, by now, students are expected to be able to generalise. The general form of the solution is considered in part d.

4 The next step is to look at where a line that is not parallel to either of the axes intersects a line that is. Students need to verify that the coordinates of the point of intersection satisfy both equations.

5 Students read off the solution to the pair of simultaneous equations, given their graphs. Again, they should be encouraged to check their answers by substitution.

6 Students now need to draw the graphs themselves. Notice that one equation has a fractional coefficient of x, so values will need to be chosen carefully. A hint is provided for this.

What do you think? 16.2A

1 Students make the link between parallel lines and a pair of simultaneous equations that do not have a solution. This can be explored further and shown to always be the case. You could extend this to look at equations that have an infinite number of solutions because they are equivalent equations.

2 illustrates the limitations of the graphical method for non-integer solutions.

Practice 16.2B

1 Students substitute $x = 0$ and $y = 0$ to find the coordinates of points where straight lines meets the axes in preparation for using this method of drawing straight lines quickly in questions later in the exercise. The phrase "cover-up method" is not used as students are encouraged to make links between the equations and the points rather than just follow a procedure that can be easily forgotten.

2 moves on to using the technique practised in question 1 to draw lines and find the points of intersection. Students should check that their answers fit both equations by substitution.

3 Students are required to form and solve a pair of simultaneous equations from the simple context of two unknown numbers. Exchanging the numbers is also explicitly considered.

4 develops this as the equations to solve are not provided.

5 shows the limitations of looking for intersections with the axes if one of the equations reduces to the form $y = ax$. Students needed to substitute a value other than 0 for x or y. Rearranging the equation could be discussed here.

6 provides sets of equations where students need to decide whether a table of values or using points of intersection with the axes is the more appropriate method for drawing the graphs.

What do you think? 16.2B

1 tests students' reasoning skills as they form equations that pass through a given point. Encourage them to consider straight lines with positive and negative gradients and to include the forms $ax + by = 0$ and $ax - by = 0$ as well as $y = mx + c$.

2 develops this further with other points, including negative x- and/or y-values. They could compare their answers to those in question 1 and note the similarities/differences.

Consolidate – do you need more?

These questions provide further practice in all the skills used in this chapter.

Stretch – can you deepen your learning?

1 explores solving simultaneous equations algebraically by substitution. Students can discuss the merits of the different approaches.

2 is another substitution example without the scaffolding provided. You could extend this by looking at other equations covered earlier in the chapter and verifying the answers.

3 is a preview of the elimination method (covered at Key Stage 4) for solving simultaneous equations that uses bar models to "subtract the equations". The bar models are provided in part a, and students should be able to construct their own for parts b and c. They could compare this method with the graphical and substitution methods.

Reflect

Students reflect on the graphical method for solving simultaneous equations and consider its limitations.

16.3 Graphs of inequalities

Small steps

- Represent inequalities

Rationale

In this short chapter, students explore a variety of representations of the solution sets of inequalities, both in one and in two variables. The initial focus is on graphical representation in keeping with the theme of the block, but number lines are explored as well. In this book, the convention is to shade the required region, rather than the region not required. The similarities between the end points of number line representations and piece-wise graphs can be noted.

Things to look out for	Interleaving opportunities
• Difficulty in deciding which side of a line to shade. Encourage students to use a simple test point (often (0, 0) will do) to see whether the value of the expression is greater than or less than the value required. • Incorrect use of circles at the ends of number line representations. Link back to piece-wise graphs.	• Forming and solving inequalities • Straight line graphs • Piece-wise graphs

Models and representations

- Graphing software can be used to illustrate regions. You will need to check that your software uses the same conventions as you teach in terms of shading, dotted lines, etc. as they are not all consistent.
- The use of number lines for showing inequalities was touched on earlier in the series, but this is the first time that students have explored it in depth. The filled circle notation for an included point links to the convention used for piece-wise graphs in Chapter 16.1.
- As always, when modelling solutions with their own graphs or reading values from graphs, students will need to be aware of the scales used on the coordinate axes.

Behind the questions

Are you ready?

Question 1 provides practice on some simple inequalities to remind students of how they are different from equations, i.e. that they have solution sets that represent an infinite range of values. Question 2 looks at substituting/testing values to prepare for testing a point in a region, and question 3 reminds students of the most basic lines that will be used when forming regions in the early part of the chapter.

Practice 16.3A

1 Students start by interpreting inequalities represented graphically. They will need to appreciate the convention that a solid line means that the values on the line are included in the region whilst a dotted line means that they are not.

2 reverses the process by asking students to sketch the regions for given inequalities using the same convention as described in question 1.

3 explores the meaning of filled-in and hollow circles in number line representations and provides some simple practice.

4 looks at regions for a line not parallel to the axes and the idea of testing a point, or points, to check which side of the line to shade. It is worth discussing the use of (0, 0), as well.

5 Students should be able to sketch these lines without tables of values, especially if they completed Chapter 16.2. Note that the point (0, 0) is not useful as a test point for part a; discuss why not.

6 builds on Practice 16.2B to quickly produce the straight line, but if students have not yet completed that chapter, there is sufficient detail in the question to enable them to access it.

7 removes the scaffolding and allows students to find for themselves some suitable points through which to draw the line.

What do you think? 16.3A

1 Students explore “overlaps” of regions.

2 illustrates closed regions, first in one and then in two variables.

3 is relatively challenging, but the line $y = x$ can be easily recognised by looking at coordinates. You can extend this by asking students to draw lines to create regions and then challenge each other to describe them. You may need to moderate their degree of complexity!

Consolidate – do you need more?

In the first question, students compare all the representations that they have met so far for showing inequalities – words, algebraic, graphical and number lines. The remainder of the exercise concentrates on interpreting and drawing the graphical representation.

Stretch – can you deepen your learning?

1 A further possible representation is introduced – set notation. Students should be familiar with union and intersection from earlier study, but may need a reminder.

2 looks at a relatively simple linear programming problem. Students need to consider what x and y represent and interpret the equations appropriately. They also need to appreciate that the solution is given by the pairs of integer values within the given region; it does not include the borders. You could extend this by looking at similar problems or vary the values in this one.

Reflect

The reflection focuses on the graphical representation, but you could ask students to discuss/record comparisons with number lines and other representations if you wish.

17.1 Handling data review

Rationale

As Book 3 does not introduce any new data handling, this chapter provides students with an opportunity to review their learning on this topic from earlier in KS3. In addition, calculating the mean from frequency tables is covered for the benefit of students who did not cover this as a Higher step in Book 2. This is extended to finding the median from frequency tables. As well as a practice exercise, the chapter provides challenges to enable students to practise their skills through projects and investigations, and thus to revisit the whole of the data handling cycle.

Reminders and links

Some examples of charts are included, along with a reminder of the key measures used when analysing data, i.e. the mean, median, mode and range. There is ample practice available on these topics, if needed, in Books 1 and 2. Alternatively, students could revise these through one of the challenge projects.

Topics covered

Finding the mode, mean and median from a frequency table is covered in the worked example. Students are introduced to the term "cumulative frequency" to help to find the median, which is formally given as the item in the $(n + \frac{1}{2})$th position. These ideas are developed in the practice exercise, in which students are asked to estimate the mean from a grouped frequency table, find the modal class and find the class in which the median lies.

Things to look out for	**Interleaving opportunities**
• Confusion between the three averages. Discuss their meanings and purpose and when each of them is or isn't appropriate. • Thinking that the range is an average. Again, discuss the purpose and use of the range for comparing sets of data. • Finding the mode or range of the frequencies rather than the data. Discuss why this error may occur.	• Data can be collected that is related to most areas of mathematics, e.g. estimating the size of angles, from probability experiments, etc. • Problems involving the calculation of the mean and other statistics can be answered by forming and solving equations.

Behind the questions

Practice 17.1A

1 This is a simple review of the basic measures and can be supported by further practice from earlier books, or by students collecting their own data, if necessary.

2 Here students practise finding the measures from a frequency table. The total frequency is 120, so students could easily represent the data as a pie chart, if desired.

3 provides fluency practice with two more frequency tables.

4 progresses to grouped frequency tables, for which students need to identify the modal class and the class in which the median lies. You may need to model the use of cumulative frequency as shown for the ungrouped frequency table in the worked example.

5 demonstrates using midpoints of class intervals to find an estimate of the mean. It might be worth discussing why this is only an estimate and how representative the midpoint is likely to be.

6 uses two more sets of data as practice at finding averages from a grouped frequency table. You could also ask students to find the mean for the table in question 4 or use tables of data from Books 1 and 2. Alternatively, students could collect their own data on which to practise these skills.

7 requires students to work in reverse to find a total from the mean and then adjust the mean given a change in the data.

8 is a slightly more complex version of question 7, requiring students to find two totals and their difference. This can be built on by combining two data sets to find an overall mean.

Challenges

1 Students explore the idea of linear interpolation to find a median. You could relate this to drawing a cumulative frequency polygon if desired, or look for a general formula using the ends of the class interval and the fraction within the interval needed to arrive at the position of the median.

2 These rich hypotheses provide opportunities for students to collect, analyse and represent data, to make conclusions and, if appropriate, to hypothesise further. There is no need to limit students to the suggested hypotheses; they can also create their own or use data from other subjects in the curriculum. These could also form the basis of discussions around the use of primary and secondary data and sampling, all of which are developed at KS4.

3 This question allows students to explore bivariate data and thus revise scatter graphs and correlation. Again, this can easily be adapted to any area of interest that can be analysed using primary or secondary data. The dangers of extrapolation can be reviewed, and it should be reiterated that correlation does not necessarily imply causality. You may wish to make use of examples of spurious correlations that can be found online to illustrate this.

17.2 Sequences review

Rationale

Sequences do feature in Book 3, in both Blocks 2 and 3, but as they are not formally revisited as a topic on their own, this chapter gives students the opportunity to review and extend their learning from Books 1 and 2. Finding an algebraic position-to-term rule was covered as a Higher step in Book 2, so this is included as a refresher or new learning depending on students' experience last year.

Reminders and links

Students are reminded of the key types of sequences and the associated vocabulary, including the fact that sequences with a constant difference between successive terms can be described both as arithmetic sequences and as linear sequences. Geometric and Fibonacci sequences, as well as the triangle, square and cube numbers are included in the reminder.

Topics covered

Term-to-term and position-to-term rules are reviewed, and the worked example models finding an expression for the nth term. This is illustrated by comparison with the multiples of the common difference, but other methods could be explored, and the expression $a + (n - 1)d$ for the nth term is used in the challenge questions. Descending sequences are included in the practice exercise.

Things to look out for	Interleaving opportunities
• Mixing up term-to-term and position-to-term rules. • Not understanding the idea of the nth term. This idea is abstract, so always relating it back to the position-to-term rule and the first term, second term, etc. is helpful. • Thinking that a sequence with common difference a has the rule $n + a$. This is easily disproved by counterexample. • Uncertainty about the sign of b in a sequence with a rule of the form $an + b$. Encourage students to check with at least two values in the sequence.	• Forming and solving equations and inequalities • Conjectures about odd/even/multiples found within sequences

Behind the questions

Practice 17.2A

1 explores sequences of the form $5n + k$, so students see that the common difference is always 5. This can be repeated with other common differences, if necessary, to reinforce the point.

2 builds directly from question 1 to identify sequences given by rules of the form $5n + k$ and to evaluate k. Listing the sequence and comparing with the 5 times table is a useful strategy.

3 Students may need guidance to identify the common difference first, and then proceed as in question 2. You could either hold a class discussion to agree on an approach or model some other sequences first.

4 Here students compare two very similar sequences. You could explore substituting $n + 1$, $n - 1$ in place of n in the rules for the sequences to deepen this comparison.

5 Only proceed to the descending sequences in this question if students are confident with ascending ones. One rule is given for students to verify in part b, as a hint to help them to find the constant on their own in the sequences in part c.

6 and **7** involve forming and solving equations and inequalities to investigate relative positions in a sequence. This may need some modelling as some students may resort to trial and improvement. These questions could be extended to looking at whether, for example, multiples of a given number lie in a sequence.

Challenges

1 looks at visual sequences where the position-to-term rule is determined by the pattern rather than just using a technique. Students should realise that the rule for the first sequence has the form $4n$ because an additional four matchsticks are added each time a new shape is formed. They are then invited to generalise to find expressions for the number of matchsticks needed to complete the first shape and the number of matchsticks added each time.

2 explores sequences given by recurrence relations. Spreadsheets may be useful for looking at patterns and whether or not sequences converge or diverge.

3 This can be illustrated by using an example sequence and substituting the first term, a, and the common difference, d, for this sequence into Flo's expression, and then expanding the brackets and simplifying. Compare the result with the nth term found in the familiar way. This can be linked to the shape patterns in question 1.

4 looks at the formal rule for geometric sequences. You could extend this by challenging students to find the rule for the nth term of a given geometric sequence.

5 In preparation for finding the nth term rule for a quadratic sequence in Key Stage 4, students explore first and second differences and compare quadratic sequences of the form $n^2 + an + b$ with n^2 to find the linear component of the rule. You could extend this either by looking at $2n^2$, $1.2n^2$, etc. (the triangle numbers are useful here) or by looking at cubic sequences to show that the third differences are constant.

6 Pascal's triangle offers a huge amount of mathematics to be explored! Students may need some initial guidance to recognise the structure, i.e. that two adjacent numbers in each row are added to give the number between these two in the row underneath:

1
1 1
1 2 1
1 3 3 1
1 4 6 4 1
1 5 10 10 5 1
1 6 15 20 15 6 1

17.3 Moving on with trigonometry

Rationale

Trigonometry is a key aspect of Key Stage 4 learning and is used to launch the Year 10 scheme of learning for White Rose Maths. This chapter can be used to provide an introduction to the topic if you wish, building on students' experience from Chapter 12.4. The focus is on identifying the sides of a right-angled triangle, and then choosing and using the correct ratio. At this stage, it is not suggested that students rote-learn the ratios through mnemonics such as SOHCAHTOA. All the formulae should be provided and displayed so that students begin to understand the process involved. Likewise, students are expected to solve the resulting equations by rearrangement rather than by the use of formula triangles. Instead, they should consider the meanings of the ratios connecting the sides.

Reminders and links

This chapter allows students to develop their knowledge of the relationships between the sides of right-angled triangles, which was first explored in the second part of Chapter 12.4, where the main focus was the 30-60-90 triangle. It may be worth revisiting this section first before progressing to more general trigonometry here.

Topics covered

In this chapter, students move on to using the sine, cosine and tangent ratios in right-angled triangles with angles of any size. They should appreciate that the ratios of the lengths of the sides are constant because the triangles are similar. Students use a calculator to find both missing sides and missing angles; they should understand that the inverse operations are denoted "$\sin^{-1} x$", etc. In the challenges section, students meet some of the ideas in more advanced trigonometry, including angles greater than 90°.

Things to look out for	Interleaving opportunities
• Incorrectly labelling the adjacent side because of confusion with the hypotenuse. Encourage students to always label the hypotenuse first. • Using calculators in the wrong mode. Always start with a simple check; for example, $\sin 30°$ should be $\frac{1}{2}$ if in degrees mode. • Trying to use trigonometry in triangles that are not right-angled. Emphasise that most ratio relationships can be used in other similar triangles, but the trigonometric ratios cannot be. Include some non-examples in your modelling. • Students may have problems with more complex rearrangement, e.g. when given $\sin 40° = \frac{12}{x}$. This is generally avoided in this chapter, but if you do explore equations of this type, link it to solving equations of the form $a = \frac{b}{x}$ using bar models.	• Solving equations • Similar triangles and enlargement • Shape and angle properties • Pythagoras' theorem

Behind the questions

Practice 17.3A

1 In this question, students use their calculators to find values of the sine, cosine and tangent ratios given the angles, and to find angles given the ratios. Students may need help to ensure that their calculators are in the correct mode and to select the right keys to find the inverses. You could discuss the similarity with squaring and finding square roots, by explaining that $\sin^{-1}(\sin x) = x$ in the same way as $\sqrt{x^2} = x$ for example.

2 provides practice at identifying the hypotenuse, opposite and adjacent sides. Encourage students to always label the hypotenuse first to ensure that this is not confused with the adjacent side.

3 is a reminder of solving equations of the form $a = \frac{b}{c}$ for each of *a*, *b* and *c*. Proficiency here is vital for success at solving equations with the trigonometric ratios, so you may need to provide intervention and further practice if there are any problems.

4 Students should follow the steps of the first worked example to find the unknown sides in the right-angled triangles. All of these reduce to the form $a = b \times \sin x/\cos x/\tan x$ to avoid any difficulties with rearrangement at this stage.

5 Likewise, students should follow the steps of the second worked example to find unknown sides in these right-angled triangles.

6 demonstrates that the unknown side can be found using either $\tan 54° = \frac{12}{x}$ and rearranging it to $x = \frac{12}{\tan 54°}$ or by using the angle sum of a triangle to find the third angle. You could add in extra examples here, including finding the hypotenuse where only rearrangement will work, if you wish.

7 Here students will need to recognise the need for using Pythagoras' theorem in triangle ABC before they can use trigonometry in triangle ADC. Discourage students from rounding in the middle of their working as this can lead to inaccuracy.

8 Students may need a hint to start by drawing a diagram. You could extend this question by looking at the effect on the size of the angle as the lengths of the sides change.

9 Again, diagrams will be useful. Students should be reminded that Pythagoras' theorem is still a valid approach when faced with problems involving right-angled triangles; it is not just trigonometry that can be used.

10 Students may need reminding that the diagonals of a rhombus intersect each other at right angles and bisect the angles at the vertices.

Challenges

1 Part a can be supported using dynamic geometry software. You could have an interesting discussion about the reason for tan 90° being undefined. Part b can be shown using any right-angled isosceles triangle.

2 This exploration of the relationships between sin, cos and tan can be scaffolded by providing a sketch with the non-right angles labelled *x* and *y* and the sides labelled *a*, *b* and *c*, to aid comparisons.

3 Finding the height of the triangle is the starting point here. Students could explore this by starting with equilateral triangles with different numerical side lengths and then generalise.

4 Students need to use directed number arithmetic to see the change in sign of the ratios. Dynamic geometry software will help them to see the change in size. You could explore trigonometric graphs as a preview of their formal study at KS4, by linking the shapes of the graphs to the values obtained from the unit circle.

5 You may have clinometers in school that students could use, or there are plenty of suggestions online as to how to make one. This will help students to see the purpose of trigonometry.